Beyond the Battle

The Evolution of Debriefing in Healthcare and Beyond

AF808541

Keith A. Beaulieu

PORTHOS
PRESS

PORTHOS
PRESS

Beyond the Battle
The Evolution of Debriefing in Healthcare and Beyond

Copyright © 2025 by Keith A. Beaulieu, Porthos Press
All Rights Reserved.

No part of this document may be reproduced or transmitted in any form or by any means, electronic, mechanical, photocopying, recording, or otherwise, without prior written permission.

Requests for permission to make copies of any part of the work should be submitted to the publisher.

Published by Keith A. Beaulieu, Porthos Press
Contact@porthospress.com

Printed in the United States of America
First Printing: May 2025

ISBN 978-1-7360792-9-4 (Paperback)

First edition, 2025
10 9 8 7 6 5 4 3 2 1

Contents

Glossary of Terms

A

AAR (After-Action Review): A structured debriefing process used in military and other fields to analyze what occurred during an event, why it happened, and how to improve.
AI (Artificial Intelligence): The simulation of human intelligence processes by machines, particularly in analyzing data during debriefings.
AR (Augmented Reality): Technology that superimposes computer-generated images on a user's view of the real world, enhancing training simulations and debriefings.

C

CISM (Critical Incident Stress Management): A set of tools and techniques designed to help individuals process traumatic events, reduce stress, and recover emotionally.
CRM (Crew Resource Management): A system originally developed in aviation to improve teamwork, communication, and decision-making, later adapted to healthcare and other fields.
CMB (Crisis Management Briefing): A large-group intervention during or after crises to provide information, alleviate panic, and connect people with support resources.

D

Debriefing: A reflective discussion conducted after an event or simulation to analyze performance, identify lessons learned, and improve future outcomes.
DASH (Debriefing Assessment for Simulation in Healthcare): A tool used to evaluate the quality and effectiveness of debriefing sessions.

E

Emotional Processing: Part of debriefing that helps participants address and reflect on their emotional responses to an event, particularly in high-stakes situations.

F

Facilitator: The person who guides a debriefing session, ensuring it is structured, effective, and psychologically safe.

G

GAS Model (Gather, Analyze, Summarize): A framework for debriefing that structures discussions into phases for collecting reactions, analyzing actions, and summarizing lessons.

H

Haptic Technology: Tools that provide tactile feedback, allowing participants to practice physical tasks in simulations with realistic touch sensations.

I

INACSL (International Nursing Association for Clinical Simulation and Learning): An organization that develops standards and best practices for simulation and debriefing in nursing education.
Interprofessional Debriefing: Reflective sessions involving team members from various disciplines to improve collaboration and communication in healthcare settings.

P

PEARLS (Promoting Excellence and Reflective Learning in Simulation): A blended debriefing framework combining learner-centered and instructor-led approaches.

R

Reaction Phase: The first phase of debriefing where participants share immediate emotional and cognitive responses to an event or simulation.

S

SSH (Society for Simulation in Healthcare): A global organization dedicated to improving simulation-based training and education in healthcare.
Simulation: A technique for practice and learning that replaces or amplifies real experiences with guided experiences, often used in training healthcare providers.

Structured Feedback: A systematic method of providing participants with observations, constructive criticism, and actionable recommendations during debriefing.

T

Team Dynamics: The interactions and behaviors among team members during a task, often analyzed in debriefings to enhance collaboration and communication.

V

Video-Assisted Debriefing: The use of recorded simulations to provide objective feedback and improve team and individual performance.

VR (Virtual Reality): An immersive technology used in simulations to replicate real-world environments for training and debriefing purposes.

Introduction

Purpose of the Book

Debriefing, once a practice rooted in military tradition, has become indispensable in healthcare simulation training. This book explores the fascinating journey of debriefing—from its origins in military operations to its pivotal role in advancing critical thinking, improving performance, and fostering effective team dynamics in healthcare settings. By delving into the history, cross-disciplinary learning, and future applications of debriefing, this resource aims to offer a comprehensive understanding of its significance in healthcare simulation programs.

The concept of debriefing is not new. In military operations, it served as a structured method to evaluate mission outcomes, extract lessons learned, and refine strategies. However, its transition into healthcare simulation represents an innovative leap, transforming how professionals are trained to navigate the complexities of patient care. Today, debriefing has become a cornerstone of simulation-based education, offering a safe and reflective space for learners to analyze their actions, identify gaps, and implement improvements. By tracing the roots of this practice and its adaptation to healthcare, this book aims to inspire simulation educators, program directors, and stakeholders to recognize the profound impact of debriefing on clinical excellence.

Importance of Debriefing

Debriefing is more than a post-event discussion; it is a powerful educational strategy that fosters critical thinking, enhances performance, and strengthens team dynamics. In healthcare simulation, debriefing allows learners to reflect on their experiences, analyze decision-making processes, and develop strategies for improvement. Research consistently highlights the benefits of debriefing in promoting learning retention, clinical competence, and collaborative practice.

Critical Thinking

One of the most significant impacts of debriefing is its ability to enhance critical thinking. Healthcare professionals often operate in high-stakes environments where decisions must be made rapidly and accurately. Debriefing allows participants to dissect scenarios, question assumptions, and consider alternative approaches. This reflective practice sharpens analytical skills, enabling professionals to think critically in real-world situations. For example, in a simulated cardiac arrest scenario, a debriefing session might focus on identifying decision points, analyzing the sequence of interventions, and exploring potential improvements in communication and execution.

Performance Improvement

Debriefing also plays a critical role in performance improvement. By identifying strengths and weaknesses, learners can develop targeted strategies to enhance their clinical skills. Feedback during debriefing sessions is instrumental in bridging the gap between theoretical knowledge and practical application. Moreover, the nonjudgmental nature of debriefing fosters an environment where participants feel comfortable acknowledging errors and embracing constructive feedback. Studies show that structured debriefing significantly improves performance outcomes, including procedural accuracy, response times, and adherence to clinical guidelines (Fanning & Gaba, 2007).

Team Dynamics

In addition to individual learning, debriefing strengthens team dynamics by promoting open communication, mutual respect, and shared accountability. Healthcare delivery often involves multidisciplinary teams working together under pressure. Effective debriefing sessions allow team members to discuss their roles, share perspectives, and address conflicts. This collaborative approach enhances trust and cohesion, ultimately improving patient outcomes. For instance, after a simulated trauma resuscitation, debriefing might highlight the importance of clear role assignments, active listening, and synchronized efforts in achieving optimal care.

Key Themes

History

The historical roots of debriefing provide a fascinating lens through which to understand its current applications. Originating in military contexts, debriefing was initially designed to evaluate missions, extract lessons learned, and improve future operations. The structured approach ensured all participants had a voice, facilitating comprehensive analyses and actionable insights. These principles laid the foundation for debriefing's adaptation to other fields, including aviation, emergency response, and healthcare simulation. By examining the evolution of debriefing, we gain valuable insights into its enduring relevance and adaptability.

Cross-Disciplinary Learning

The transition of debriefing from military to healthcare settings exemplifies the power of cross-disciplinary learning. Borrowing best practices from diverse domains has enriched the field of healthcare simulation, creating innovative methods to address complex challenges. For instance, techniques such as the "After Action Review" from military settings and "Crew Resource Management" from aviation have been seamlessly integrated into debriefing frameworks. These cross-disciplinary influences have enhanced the effectiveness of debriefing in promoting learning, safety, and teamwork.

Future Applications

As healthcare continues to evolve, so does the potential for debriefing to address emerging challenges. Advances in technology, such as virtual reality and artificial intelligence, offer exciting opportunities to revolutionize debriefing practices. Simulated environments can now provide real-time data, enabling more precise analyses and personalized feedback. Furthermore, the principles of debriefing are increasingly being applied beyond simulation, influencing quality improvement initiatives, incident reviews, and organizational learning. By exploring these future applications, this book highlights the transformative potential of debriefing in shaping the future of healthcare education and practice.

Structure of the Book

This book is organized into several sections, each designed to provide a comprehensive understanding of debriefing in healthcare simulation:

1. **Historical Foundations:** An exploration of the military origins of debriefing and its journey into healthcare.
2. **Core Principles and Methodologies:** A detailed examination of debriefing frameworks, techniques, and best practices.
3. **Impact on Learning and Performance:** An analysis of the evidence supporting debriefing's role in enhancing critical thinking, clinical skills, and teamwork.
4. **Cross-Disciplinary Insights:** A discussion of how lessons from other industries have shaped debriefing practices in healthcare simulation.
5. **Innovations and Future Directions:** Exploring emerging trends, technologies, and applications in debriefing.

Conclusion

The journey of debriefing from its military roots to its current role in healthcare simulation is a testament to its adaptability and impact. By examining its history, principles, and applications, this book aims to inspire a deeper appreciation for debriefing as a learning, improvement, and collaboration tool. Whether you are a simulation educator, program director, or healthcare professional, this resource offers valuable insights and practical strategies to enhance your practice. Together, let us unlock the full potential of debriefing to transform healthcare education and improve patient outcomes.

Chapter 1
Concept of Debriefing in Warfare

Early Use in Military Strategy

The debriefing practice can be traced back to ancient warfare, which was used informally during war councils and post-battle discussions. These gatherings were crucial for assessing the outcomes of battles, understanding enemy tactics, and refining strategies for future engagements. Leaders like Sun Tzu, author of *The Art of War*, emphasized the importance of reflection and analysis in achieving victory. Sun Tzu argued that understanding the dynamics of both successes and failures was critical to adapting strategies, a principle echoed in his famous quote, "Know your enemy and know yourself and you can fight a hundred battles without disaster" (Giles, 1910).

In ancient Rome, commanders would convene with their officers after battles to discuss troop performance, assess the effectiveness of formations, and adapt strategies. These discussions often included successes and failures, offering a balanced perspective on the engagement. Similarly, medieval war councils brought together military leaders to review battlefield outcomes and make decisions about subsequent campaigns. The Knights Templar, for example, were known for their disciplined post-battle reviews, dissecting tactical errors and sharing lessons among their ranks (Barber, 1994). These early debriefing forms were driven by the need to extract actionable lessons from every encounter, emphasizing the value of learning as a continuous process.

Formalization in Modern Military Practices

The evolution of debriefing into a formalized practice occurred alongside the professionalization of modern militaries. By the 19th and 20th centuries, military organizations recognized the importance of structured post-mission reviews to enhance operational effectiveness. This shift was particularly evident during World War II, when the United States and Allied forces institutionalized debriefing as a critical component of mission planning and execution (Ambrose, 1997). For instance, B-17 bomber pilots conducted detailed debriefings, called "**Interrogations**," after every mission. During these sessions, crews replayed the mission from each position within the bomber, providing a comprehensive account of their experiences. Leadership and intelligence teams meticulously recorded information on enemy strength and type, the target status, mission results, and the positions and locations of downed ally aircraft. These debriefings informed future operations and contributed to the strategic understanding of enemy movements and tactics.

The military continues to employ the "**Hot Wash**" as an immediate, informal debriefing conducted directly after an exercise, drill, or emergency. The primary objective is to quickly capture facts and lessons learned and identify areas for improvement while the insights remain fresh. This rapid feedback loop can enhance future performance and even save lives in high-stakes environments. Following the Hot Wash, a more formal debriefing, known as the "**After Action Review**" (AAR), typically occurs in a structured setting. Developed by the U.S. Army in the 1970s, the AAR provides a systematic framework for analyzing operations, identifying lessons learned, and improving future performance. Unlike traditional methods focused solely on outcomes, the AAR emphasizes process-oriented learning, encouraging participants to evaluate their actions, decisions, and team dynamics. This dual approach to debriefing—Hot Wash followed by an AAR—has become a cornerstone of continuous improvement in military and other high-risk operations.

Hot Wash	After-Action Review
• Informal and immediate • Individuals directed to attend • Open and agile discussion format • Gather immediate observations	• Formal and structured • Key stakeholders • Discussion follows framework. • Identify strengths, weaknesses, opportunities, and threats

• Designed for brainstorming and collaboration	• Designed for organizational improvement

How to Conduct a Hot Wash

Effective Hot Wash requires a clear structure to gather meaningful insights without delay. The following steps outline the process, as adapted from best practices:

- **Gather the Team**: Immediately following the event, assemble all participants, ensuring everyone involved has the opportunity to contribute.
- **Set Expectations**: Define the purpose of the Hot Wash—to identify successes, challenges, and areas for improvement. Emphasize the need for honesty and constructive feedback.
- **Facilitate Open Communication**: Use targeted questions to guide discussion, such as:
  - What went well during the exercise or operation?
  - What didn't go as planned?
  - What could be done differently next time?
- **Focus on Key Takeaways**: Prioritize the most critical observations and ensure they are documented.
- **Assign Follow-Up Tasks**: If immediate action items arise, assign team members responsibility for addressing them promptly.

Example Table for a Hot Wash

Component	Description	Example Questions
What Worked Well	Identify successes to replicate in the future.	"What strategies were most effective?"
What Didn't Work	Highlight challenges or failures that need addressing.	"What obstacles did we encounter?"
Recommendations	Propose solutions to improve future outcomes.	"What could we change to avoid these challenges?"
Action Items	Assign responsibilities for follow-up.	"Who will implement the suggested changes?"

Why the Hot Wash is Important

By conducting a Hot Wash immediately after an event, teams can capture front-of-mind insights often lost as time passes. This swift debriefing sets the stage for a more in-depth analysis of the AAR. These processes create a feedback loop that fosters continuous learning and operational improvement, whether on the battlefield or in other high-stakes environments.

During the Cold War, debriefing became standard practice in intelligence and reconnaissance missions. Spies and operatives were debriefed to extract valuable information about enemy activities, operational challenges, and potential risks. These debriefings were meticulously documented, ensuring that lessons learned could inform future operations. The emphasis on precision, thoroughness, and actionable insights highlighted the strategic value of debriefing in complex, high-stakes environments (Richelson, 1997).

Objectives of Military Debriefing

The primary objectives of military debriefing have remained consistent over time: improving strategy, identifying errors, and fostering team cohesion. Each of these objectives plays a vital role in enhancing the effectiveness and resilience of military organizations.

Improving Strategy

Debriefing provides a structured process for analyzing mission outcomes and refining strategies. By reviewing what worked and what didn't, military leaders can develop more effective plans for future operations. This iterative approach to strategy development ensures that lessons learned are systematically integrated into decision-making processes. For example, after a failed assault, commanders might use debriefing sessions to identify weaknesses in their approach, such as inadequate reconnaissance or poor communication, and implement corrective measures (Dorn, 2016). The ability to pivot and adapt strategies in response to debriefing findings has been credited with turning the tide in numerous conflicts, including key campaigns during World War II.

Identifying Errors

Mistakes are inevitable in any complex operation but represent valuable learning opportunities. Military debriefing focuses on identifying errors, understanding their root causes, and developing strategies to prevent recurrence. This emphasis on continuous improvement has been instrumental in reducing operational risks and enhancing mission success rates. For instance, debriefing sessions after training exercises often highlight procedural errors, such as improper weapon handling or failure to follow standard operating procedures. This enables participants to address these issues before real-world deployments (Fanning & Gaba, 2007).

Fostering Team Cohesion

Teamwork is at the heart of military operations, and debriefing is a powerful tool for fostering cohesion among team members. By providing a platform for open communication and mutual feedback, debriefing sessions help build trust, resolve conflicts, and strengthen interpersonal relationships. This collaborative environment encourages individuals to take ownership of their actions and support one another in achieving shared goals. In high-stakes scenarios, such as special forces missions, the ability to operate as a cohesive unit can mean the difference between success and failure (Gerras, 2008).

Debriefings Are Not...

- A punitive process
- A stress management/therapy session
- A search to focus on negative emotions or stress
- A casual conversation
- A lecture
- An evaluation or critique
- A formal investigation
- A search for scapegoats

Key Takeaways

1. Debriefing has deep historical roots, originating in ancient war councils and evolving into a structured practice in modern militaries.
2. Formalized approaches, such as the After-Action Review, emphasize process-oriented learning and have become a cornerstone of military strategy.
3. Military debriefing's primary objectives include improving strategy, identifying errors, and fostering team cohesion, all of which contribute to operational effectiveness.

4. Lessons learned from military debriefing have influenced other fields, including healthcare, aviation, and emergency response.

Reflection

The evolution of debriefing demonstrates its enduring relevance and adaptability across centuries and disciplines. By examining its historical and modern applications, we gain valuable insights into the universal learning principles from experience. Reflecting on how military practices like the After-Action Review prioritize open communication and constructive feedback, we can see the potential for these methods to enhance collaboration and performance in diverse settings. As you continue through this book, consider how these lessons might be applied in your own professional or organizational contexts to drive improvement and innovation.

Conclusion

The concept of debriefing in warfare has evolved from informal discussions in ancient war councils to the structured practices of modern militaries. By improving strategy, identifying errors, and fostering team cohesion, debriefing has become an indispensable tool for enhancing operational effectiveness. The lessons learned from military debriefing have shaped the development of professional armed forces and influenced other fields, including healthcare, aviation, and emergency response. In the following chapters, we will explore how the principles of military debriefing have been adapted and applied to healthcare simulation, creating innovative approaches to education and performance improvement.

References

Ambrose, S. E. (1997). Citizen Soldiers: The U.S. Army from the Normandy Beaches to the Bulge to the Surrender of Germany. Simon & Schuster.

Barber, M. (1994). The New Knighthood: A History of the Order of the Temple. Cambridge University Press.

Dorn, E. (2016). Winning by the Numbers: How After Action Reviews Improve Performance. Harvard Business Review.

Fanning, R. M., & Gaba, D. M. (2007). The role of debriefing in simulation-based learning. Simulation in Healthcare, 2(2), 115-125.

Gerras, S. J. (2008). Team Building in the Military: Lessons Learned from Debriefing Practices. Military Review.

Giles, L. (1910). The Art of War by Sun Tzu. Clarendon Press.

Richelson, J. T. (1997). The U.S. Intelligence Community. Westview Press.

U.S. Army. (1984). After Action Reviews: Guidelines and Techniques. U.S. Government Printing Office.

Chapter 2
Evolution of Military Debriefing Practices

Development During World War II: The Need for Systemic Learning from Combat Experiences

World War II's sheer scale and complexity necessitated a systematic approach to learning from combat experiences. Military leaders quickly realized that the ability to adapt tactics and strategies based on real-time feedback could mean the difference between success and failure. This era saw the formalization of debriefing practices as essential tools for operational improvement.

For instance, bomber crews, particularly those flying B-17s, engaged in detailed debriefing sessions called "interrogations." These sessions replayed the mission from every crew member's perspective, capturing a comprehensive account of events. Intelligence officers documented critical information, such as enemy strength and type, target status, mission results, and the locations of downed ally aircraft. These debriefings often painted vivid pictures of the chaos and precision of aerial warfare, bringing to life the harrowing experiences of crew members. The meticulous documentation informed immediate tactical adjustments and contributed to long-term strategic planning (Ambrose, 1997).

The Allied forces also used these debriefings to refine bombing strategies, improve navigational techniques, and enhance coordination among air, land, and sea units. Pilots frequently described the tension of flying through enemy flak or navigating by starlight, while gunners recounted the split-second decisions that could save or doom a mission. The feedback loop created by these debriefings ensured that lessons learned from one mission could be applied to future operations, significantly increasing operational effectiveness.

Vietnam War and Psychological Debriefing: Addressing Trauma and Performance Improvement

The Vietnam War introduced a new dimension to debriefing practices: addressing the psychological toll of combat. The unconventional nature of the conflict, coupled with prolonged exposure to high-intensity combat, necessitated an emphasis on both operational and emotional debriefing.

Military psychologists began incorporating techniques to help soldiers process traumatic experiences. After intense firefights or ambushes, soldiers would gather in small groups to share their experiences. These sessions were often raw and emotional, with participants recounting moments of terror, courage, and loss. These debriefings aimed to reduce immediate psychological stress and prevent long-term conditions like post-traumatic stress disorder (PTSD).

In addition to emotional support, these sessions also focused on operational feedback, allowing soldiers to identify what worked and what didn't in their engagements. Commanders often emphasized the need to learn from successes and mistakes, fostering a culture of resilience and adaptability. This dual-purpose approach highlighted the evolving understanding of debriefing as not just a tool for tactical learning but also for preserving the mental well-being of personnel (Shay, 1994).

The psychological debriefings of the Vietnam era laid the groundwork for modern practices in critical incident stress management (CISM). These practices are now widely used in both military and civilian contexts, including healthcare and emergency response, where the psychological impact of critical events can be equally profound.

Case Studies of Successful Military Debriefing

Operation Overlord: D-Day and the Power of Feedback

The success of Operation Overlord, the Allied invasion of Normandy in 1944, owes much to the rigorous debriefing practices that preceded and followed the mission. Prior to D-Day, extensive rehearsals were conducted, each followed by detailed debriefings. Soldiers practiced storming beaches under simulated fire while commanders reviewed their strategies and adjusted plans based on feedback from these exercises. These sessions identified coordination, communication, and logistics gaps, leading to critical adjustments before the invasion.

Post-mission debriefings were equally thorough. Soldiers recounted their experiences navigating heavily fortified beaches, while engineers shared the challenges of clearing obstacles under fire. Intelligence teams combined these firsthand accounts to build a comprehensive battlefield picture. This feedback informed the subsequent phases of the invasion and provided lessons that influenced Allied strategies for the remainder of the war (Ryan, 1959).

Desert Storm: Integration of Technology and Debriefing

During the Gulf War, integrating advanced technologies, such as real-time satellite imagery and digital communications, enhanced debriefing practices. After each operation, commanders and intelligence officers used these tools to conduct highly detailed debriefings, analyzing battlefield outcomes with unprecedented accuracy.

One notable example was the debriefing process for air sorties. Pilots reviewed mission footage and telemetry data alongside intelligence teams, enabling precise evaluations of target engagement, mission execution, and enemy response. These sessions often included vivid accounts of split-second decisions made in combat, such as evasive maneuvers to avoid surface-to-air missiles. These technologically augmented debriefings provided previously unattainable detail, setting a new standard for post-mission analysis (Atkinson, 1993).

Special Forces Operations: The Role of After-Action Reviews

Special Forces units, known for their high-stakes and complex missions, have long relied on debriefing practices to refine their performance. The use of **After-Action Reviews** (AARs) has been particularly impactful. These structured sessions allow team members to candidly discuss what went well, what didn't, and how to improve. Emphasizing transparency and accountability has fostered a culture of continuous learning within Special Forces communities (U.S. Army, 1984).

For example, during counterterrorism operations in Afghanistan, Special Forces teams used AARs to adapt their tactics to the challenges of asymmetric warfare. Operators described the complexities of urban combat, including the need to distinguish civilians from combatants in split-second decisions. Feedback from these debriefings led to innovations in urban combat techniques, intelligence gathering, and coordination with allied forces, significantly enhancing mission success rates.

After-Action Review (AAR): A Framework for Learning and Improvement

The After-Action Review (AAR) is a hallmark of military debriefing, designed to analyze operations in a structured and participatory manner. Developed in the 1970s by the U.S. Army, the AAR provides a consistent framework for evaluating performance, promoting accountability, and fostering continuous improvement. Its structured approach encourages reflective learning and collaborative problem-solving, making it highly adaptable to healthcare settings.

Key Components of an AAR:

What was supposed to happen?
Establishing the objectives and expectations of the activity.

What actually happened?
Comparing the actual outcomes with intended goals.

What went well?
Highlighting successes and reinforcing effective practices.

What needs improvement?

Identifying gaps, errors, and areas for growth.

How can we improve?
Developing actionable strategies for future implementation.

Benefits of the AAR Framework:

- **Systematic Analysis**: Encourages teams to dissect events methodically, ensuring no critical detail is overlooked.
- **Collaboration**: Promotes open dialogue and shared accountability, enhancing team cohesion.
- **Actionable Insights**: Focuses on generating specific, practical recommendations for future scenarios.

Challenges of the AAR Framework

- Creating a safe, non-punitive environment for honest feedback
- Ensuring insights leads to actionable change
- Overcoming time constraints
- Maintaining consistency

Adapting AAR to Healthcare Simulation: In healthcare simulation, AAR can be integrated to evaluate complex clinical scenarios, improve decision-making, and enhance interprofessional collaboration. For example, after a simulated cardiac arrest, participants can use the AAR structure to analyze their adherence to advanced life support protocols, communication dynamics, and response times.

Key Takeaways

- The systemic learning approaches developed during World War II established a foundation for modern debriefing practices, particularly through detailed mission reviews like those conducted by B-17 bomber crews.
- The Vietnam War highlighted the psychological dimensions of debriefing, addressing trauma, and promoting resilience alongside operational learning.
- Technological advancements, such as real-time data and satellite imagery, have enhanced the depth and precision of debriefing practices, as demonstrated during the Gulf War.

- Case studies, including Operation Overlord and Special Forces AARs, underscore structured feedback's strategic and operational value in achieving mission success.
- The evolution of debriefing practices reflects a continuous adaptation to meet the changing demands of warfare, offering lessons that extend beyond the military to other disciplines.

Reflection

The journey of military debriefing practices reveals the critical importance of learning from experience in high-stakes environments. From the methodical mission analyses of World War II to integrating psychological and technological insights in later conflicts, debriefing has proven to be a dynamic and versatile tool. Reflecting on these practices prompts us to consider how structured feedback can drive improvement in military contexts and other high-pressure fields, such as healthcare and emergency response. By fostering open communication, accountability, and a commitment to learning, debriefing can transform challenges into opportunities for growth and innovation.

Conclusion

The evolution of military debriefing practices reflects the growing recognition of their strategic, operational, and psychological value. From the systemic learning necessitated by World War II to the dual focus on performance and mental health during the Vietnam War, debriefing has continuously adapted to meet the needs of modern warfare. Case studies, such as those from Operation Overlord, Desert Storm, and Special Forces operations, demonstrate the transformative impact of effective debriefing on mission success.

As the next chapters will explore, the principles and methodologies developed in military contexts have profoundly influenced other fields, particularly healthcare simulation. By understanding the evolution of military debriefing, we can better appreciate its enduring relevance and potential for cross-disciplinary application.

References

Ambrose, S. E. (1997). Citizen Soldiers: The U.S. Army from the Normandy Beaches to the Bulge to the Surrender of Germany. Simon & Schuster.

Atkinson, R. (1993). Crusade: The Untold Story of the Persian Gulf War. Houghton Mifflin.

Ryan, C. (1959). The Longest Day: June 6, 1944. Simon & Schuster.

Shay, J. (1994). Achilles in Vietnam: Combat Trauma and the Undoing of Character. Scribner.

U.S. Army. (1984). After Action Reviews: Guidelines and Techniques. U.S. Government Printing Office.

Chapter 3
Core Principles of Military Debriefing

Structured Feedback and Reflection

One of the fundamental principles of military debriefing is the emphasis on structured feedback and reflection. Military debriefing sessions are meticulously planned to ensure that every participant's perspective is considered and that actionable insights are drawn from the experience. Structured feedback is the backbone of these sessions, providing a clear framework for participants to evaluate their actions, decisions, and team dynamics.

A typical structured debrief follows a sequence: participants recount what occurred, analyze why it happened, and discuss how to improve future performance. This organized approach ensures that the focus remains on learning rather than blame. For instance, after a simulated urban combat exercise, a debriefing might follow a framework like the After-Action Review (AAR), where participants address key questions: "What was supposed to happen?" "What actually happened?" and "What can we do differently next time?" (U.S. Army, 1984). These systematic discussions encourage critical thinking and collective problem-solving.

Structured reflection also plays a vital role in reinforcing learning. By encouraging participants to revisit key moments from an operation, debriefing sessions help crystallize lessons that might otherwise be forgotten. In one notable instance, a military unit conducting a training exercise on night operations discovered, through structured reflection, that their reliance on a single point of failure - specific piece of equipment—was a critical risk. Through the debrief, they identified alternate contingencies, which later proved vital during actual missions.

This reflective process is particularly valuable in high-stakes environments where immediate operational demands often overshadow opportunities for introspection. Over time, this principle of structured feedback and reflection fosters a culture of continuous improvement and adaptability.

Focus on Learning Rather than Punishment

Another core principle of military debriefing is the commitment to fostering a learning-oriented environment. Military operations often involve high levels of stress, risk, and uncertainty. In such contexts, mistakes are inevitable, but how those mistakes are addressed determines whether they lead to growth or undermine morale.

Effective debriefing shifts the focus from punishment to learning. By creating a safe space where individuals can openly discuss errors without fear of retribution, debriefing sessions encourage honest and constructive dialogue. For example, during debriefings in Special Forces units, leaders often emphasize that failures are shared responsibilities, reflecting the collective nature of military operations. This approach facilitates a deeper understanding of errors and strengthens team cohesion.

This principle was demonstrated during a joint military exercise involving multiple allied forces. A navigational error led to significant delays in mission execution, but instead of assigning blame, the debriefing focused on understanding the root cause of the error: inadequate pre-mission coordination between teams. By addressing the systemic issue, participants ensured that future collaborations would be more seamless.

The focus on learning rather than punishment is particularly evident in modern After-Action Reviews. Participants are encouraged to frame mistakes as opportunities to identify gaps and improve processes. This mindset aligns with the broader goal of military debriefing: to enhance operational effectiveness through continuous learning. Military organizations can cultivate resilience and innovation by prioritizing growth over blame, even in the face of adversity.

Emotional Processing and Psychological Safety

Military debriefing also recognizes the critical importance of emotional processing and psychological safety. Combat and other high-intensity operations often expose personnel to significant psychological stress. Addressing these emotional experiences is essential not only for individual well-being but also for maintaining operational readiness.

Emotional processing is integrated into many debriefing practices, particularly in contexts where personnel have faced traumatic events. For example, critical incident debriefings often include opportunities for participants to share their feelings, acknowledge their experiences, and receive support from peers and leaders. These sessions frequently include moments of profound connection, such as a soldier recounting the loss of a comrade and being met with solidarity from their unit. This process helps alleviate the immediate emotional impact of stressful events and reduces the risk of long-term psychological conditions, such as post-traumatic stress disorder (Shay, 1994).

Psychological safety is a cornerstone of effective debriefing. By creating an environment where participants feel secure in expressing their thoughts and emotions, debriefing sessions promote openness and trust. Leaders play a crucial role in fostering this safety by modeling vulnerability, validating participants' experiences, and reframing challenges as opportunities for growth. For instance, after a challenging mission, leaders might share their reflections on what they learned, demonstrating that even experienced personnel can grow from feedback.

A notable example comes from medical evacuation units in combat zones, where teams often face the emotional toll of caring for critically wounded soldiers. During debriefings, team members are encouraged to express their feelings about particularly difficult cases, creating a space to collectively process their experiences. This practice strengthens individual resilience and enhances the team's ability to function effectively under pressure.

Key Takeaways

- Structured feedback and reflection are essential components of military debriefing, ensuring that lessons are drawn systematically and applied effectively.
- A focus on learning rather than punishment fosters an open and constructive environment, enabling teams to address mistakes without fear and prioritize growth.
- Addressing emotional processing and ensuring psychological safety are critical for maintaining the mental well-being of personnel and enhancing team cohesion.
- The principles of military debriefing—structure, learning, and emotional support—create a foundation for continuous improvement and resilience in high-stakes environments.

Reflection

The principles of military debriefing offer profound lessons for any organization or individual seeking growth and improvement. Structured feedback ensures that insights are actionable and targeted, while a focus on learning over punishment cultivates trust and openness. Perhaps most importantly, recognizing the emotional dimension of experiences underscores the importance of psychological safety in fostering collaboration and resilience. Reflecting on these practices invites us to consider how we might adopt similar principles in our contexts to create environments where individuals and teams can learn, adapt, and thrive.

Whether it is a military operation or a civilian initiative, these principles remind us of the power of learning from experience. They challenge us to embrace feedback, approach mistakes with curiosity rather than judgment, and prioritize the well-being of our teams. Doing so creates the conditions necessary for sustained success and innovation.

Conclusion

The core principles of military debriefing—structured feedback and reflection, a focus on learning rather than punishment, and emotional processing with psychological safety—form the foundation of its effectiveness. These principles ensure that debriefing serves as a powerful tool for learning, improvement, and resilience. Military debriefing practices create environments where individuals and teams can thrive even in the most demanding circumstances by emphasizing structure, fostering a growth-oriented mindset, and addressing the emotional dimensions of operations.

As this book continues, we will explore how these principles have been adapted and applied to other fields, such as healthcare simulation and emergency response, highlighting their universal relevance and transformative potential.

References

Shay, J. (1994). Achilles in Vietnam: Combat Trauma and the Undoing of Character. Scribner.

U.S. Army. (1984). After Action Reviews: Guidelines and Techniques. U.S. Government Printing Office.

Figure 1 Example of a military style prebrief.
Photograph courtesy of Steven McCloud, U.S. Southern Command Public Affairs

Example of a military-style debrief.
Photograph courtesy of Spc. Sara Stalvey, USA, 55th Signal Company

Example of military-style debrief.
Photograph courtesy of Department of Defense

Chapter 4
Transition to Civilian Sectors

Adoption of Military Techniques in Aviation

Adopting military debriefing practices in the aviation industry set the stage for groundbreaking advancements in safety, competency, and mastery through simulation and Crew Resource Management (CRM). Aviation's inherently high-risk and high-pressure environment required systematic methods to enhance technical and non-technical skills, creating a fertile ground for integrating simulation-based training and military-inspired practices.

Military Influence and the Development of CRM

The transition of structured debriefing and After-Action Reviews (AARs) from military to civilian aviation began in the mid-20th century. The U.S. Navy, known for its rigorous debriefing practices among military pilots, provided a model for the aviation industry. Military pilots emphasized the importance of analyzing technical aspects of their missions and human factors such as communication, decision-making, and stress management (Helmreich & Merritt, 1998). These principles formed the foundation for Crew Resource Management (CRM), a systematic approach to improving teamwork and communication among flight crews.

Key Components of CRM

CRM is a multidisciplinary framework designed to address human factors in aviation. It focuses on improving team dynamics, reducing human error, and fostering a safety-first culture. Key components of CRM include:

Structured Communication

CRM emphasizes clear and standardized communication to minimize ambiguity. For example, flight crews are trained to use standardized phraseology, such as "confirm clearance for takeoff" rather than informal or ambiguous language, to ensure mutual understanding between team members and air traffic control. Structured communication also involves active listening and the confirmation of received instructions, which helps reduce the risk of miscommunication.

Leadership and Followership

Effective leadership in the cockpit fosters a collaborative and respectful environment. CRM encourages captains to practice inclusive leadership, ensuring all crew members feel comfortable voicing concerns. Conversely, followership involves subordinate crew members being assertive when identifying potential safety threats, regardless of rank or experience.

Situational Awareness

Maintaining situational awareness is critical for anticipating potential risks and understanding the operational environment. CRM equips crews with tools to monitor flight conditions, recognize deviations, and proactively adjust to maintain safety.

Decision-Making

CRM trains crews to apply systematic decision-making processes, particularly during high-stress scenarios. These processes involve gathering and analyzing information, evaluating options, and implementing actions while cross-checking decisions against available data. This reduces impulsivity and helps crews mitigate the risks of cognitive biases.

Teamwork and Collaboration

Teamwork is at the heart of CRM. It emphasizes trust, respect, and interdependence among crew members. Crews are trained to leverage the team's collective knowledge, ensuring input is considered from all members, including flight attendants and air traffic controllers, when making critical decisions.

Error Management

Recognizing that human error is inevitable, CRM focuses on error management strategies, including error detection, mitigation, and recovery. These strategies aim to minimize the impact of errors and prevent them from escalating into catastrophic outcomes.

Aviation Disasters and CRM

Tenerife Disaster (1977)

The 1977 Tenerife airport disaster remains the deadliest accident in aviation history, resulting in 583 fatalities and reshaping the industry's approach to safety and communication. On March 27, 1977, two Boeing 747s—a KLM aircraft and a Pan Am aircraft—collided on the runway at Los Rodeos Airport (now Tenerife North Airport) in the Canary Islands. This tragedy was caused by miscommunication, procedural lapses, and human error, compounded by adverse weather conditions and airport congestion.

Photograph in the common domain

Contributing Factors to the Disaster

- **Dense Fog and Limited Visibility**
  - Heavy fog on the day of the crash reduced visibility to nearly zero, making it difficult for pilots to see other aircraft on the runway. The airport's lack of ground radar further exacerbated the situation, as air traffic controllers relied solely on radio communication and pilot reports to manage aircraft movements. This critical limitation created an environment where miscommunication had catastrophic potential.

- **Airport Congestion**
  - Los Rodeos Airport was not designed to handle the traffic volume it experienced that day. A nearby bombing at Gran Canaria Airport had forced numerous flights, including the KLM and Pan Am 747s, to divert to Los Rodeos. The airport's

single runway and limited taxiway infrastructure led to significant delays and operational confusion as aircraft were rerouted and repositioned.

- **Miscommunication Between Pilots and Air Traffic Control**
  - Ambiguity in radio communication played a central role in the disaster. The KLM captain, eager to depart after a long delay, misunderstood an ambiguous clearance from air traffic control as permission to take off. Meanwhile, the Pan Am 747, still taxiing on the runway, was obscured by fog and unaware of the KLM aircraft's position. Using non-standard phraseology, such as "We are at takeoff," created confusion about the KLM aircraft's intentions.

- **Hierarchical Dynamics in the Cockpit**
  - Investigators found that the KLM captain, an experienced and highly respected pilot, held significant authority in the cockpit. This dynamic likely inhibited the first officer and flight engineer from questioning his decision to initiate takeoff. Despite their apparent awareness of the ambiguity in air traffic control's instructions, they did not challenge the captain's actions—a classic example of what is now called the "authority gradient."

- **Lack of Standardized Protocols**
  - At the time, there were no universally enforced protocols to ensure mutual understanding between pilots and air traffic controllers. Terms such as "ready for takeoff" or "at takeoff" were not standardized, leaving room for interpretation. This lack of consistency and lack of formalized cross-checking procedures within the cockpit created a perfect storm for disaster.

Sequence of Events

The KLM aircraft, having completed refueling, taxied to the runway and lined up for departure. The Pan Am aircraft, directed to taxi down the same runway and exit at a designated taxiway, became disoriented in the fog and missed the assigned turnoff.

At 17:06, the KLM captain informed air traffic control, "We are now at takeoff," which the crew interpreted as a declaration rather than a question. Simultaneously, the Pan Am crew reported, "We are still taxiing on the runway." However, due to overlapping radio transmissions, this message was not clearly received by the KLM crew or air traffic control.

The KLM aircraft began its takeoff roll at full thrust, rapidly accelerating toward the unseen Pan Am jet. By the time the Pan Am crew spotted the KLM aircraft's landing lights emerging from the fog, it was too late to avoid a collision. Despite an attempt by the KLM captain to rotate and lift off early, the aircraft's tail struck the runway, and its fuselage sheared through the Pan Am jet's upper cabin. Both aircraft burst into flames, killing 583 people.

Lessons Learned

The Tenerife disaster exposed critical vulnerabilities in aviation operations and highlighted the dire consequences of poor communication and hierarchical cockpit dynamics. Among the key lessons were:

1. **The Need for Standardized Communication**
   The ambiguity in radio exchanges between the KLM crew, the Pan Am crew, and air traffic control underscored the necessity of precise, standardized phraseology. Phrases such as "cleared for takeoff" and "hold position" were later mandated to eliminate potential misinterpretation.

2. **Importance of Assertiveness in the Cockpit**
   The KLM first officer and flight engineer's failure to challenge their captain's decision revealed the dangers of excessive difference to authority. CRM training now emphasizes assertiveness and the responsibility of all crew members to voice concerns, regardless of rank.

3. **Introduction of CRM**
   The Tenerife disaster catalyzed the development and widespread adoption of Crew Resource Management (CRM). CRM focused on communication, teamwork, and decision-making to address many of the human factors that contributed to the tragedy.

Broader Impact on Aviation Safety

The Tenerife disaster was a wake-up call for the aviation industry, leading to sweeping changes in safety protocols and training programs. CRM has become a global standard in the decades since, with its principles integrated into every aspect of aviation operations. The tragedy also underscored the need for constant vigilance, reminding the industry that even the most experienced professionals are not immune to the risks of human error.

While the loss of 583 lives was devastating, the lessons learned from the Tenerife disaster have undoubtedly saved countless others. Its legacy is a testament to the importance of continuous improvement and the shared responsibility of all stakeholders to prioritize safety above all else.

ValuJet Flight 592 (1996)

On May 11, 1996, ValuJet Flight 592, a McDonnell Douglas DC-9, crashed into the Florida Everglades shortly after takeoff from Miami International Airport, killing all 110 passengers and crew onboard. This devastating accident exposed serious lapses in safety oversight, maintenance practices, and emergency response protocols, providing valuable lessons for the aviation industry.

The Sequence of Events

ValuJet Flight 592 was scheduled to fly from Miami to Atlanta. Shortly after takeoff, a fire broke out in the cargo hold due to improperly stored oxygen generators. These generators, which produce oxygen as a byproduct, were expired, improperly labeled, and lacked protective safety caps, making them highly flammable. The intense fire quickly spread, disabling critical systems, filling the cabin with smoke, and overwhelming the flight crew.

Photograph courtesy of the Atlanta Journal-Constitution

Within just ten minutes of takeoff, the crew lost control of the aircraft as it succumbed to the fire and structural failure, crashing into the remote

Everglades. The speed and intensity of the fire, coupled with a lack of effective containment measures, made survival impossible.

Contributing Factors to the Crash

1. **Improper Handling of Hazardous Materials**

   The primary cause of the crash was the transportation of oxygen generators in the cargo hold, which violated Federal Aviation Administration (FAA) regulations. These generators were not classified as hazardous materials, leading to their improper handling and loading by SabreTech, the airline's maintenance contractor. The lack of clear labeling and safety caps on the generators allowed them to ignite during the flight.

2. **Lapses in Maintenance Oversight**

   SabreTech, responsible for maintaining and shipping the oxygen generators, failed to follow proper safety protocols. Investigations revealed that SabreTech employees knowingly loaded the generators onto the aircraft without understanding their hazardous nature. ValuJet's reliance on third-party contractors for maintenance further highlighted a lack of oversight and accountability within its operations.

3. **Deficiencies in Emergency Response Training**

   The flight crew struggled to manage the escalating crisis due to the rapid spread of the fire and the cabin's limited visibility. Although trained in emergency procedures, the severity of the fire overwhelmed their ability to implement standard protocols. This highlighted the need for improved training to handle extreme and unexpected scenarios.

4. **Regulatory Weaknesses**

   At the time, ValuJet was known for its aggressive cost-cutting strategies as an ultra-low-cost carrier, leading to concerns about its safety practices. The FAA had previously flagged ValuJet for safety violations, but the airline continued operations with minimal regulatory intervention. The crash underscored the FAA's failure to enforce stricter oversight, particularly for low-cost carriers reliant on outsourced maintenance.

CRM and ValuJet Flight 592

While the crash primarily stemmed from maintenance and regulatory issues, the flight crew's inability to effectively communicate and manage the emergency emphasized the importance of CRM principles:

1. **Communication During Crisis**

   The fire's rapid progression made communication between the cockpit and cabin crew critical. However, smoke and failing systems hindered the crew's situational awareness, complicating their response coordination. CRM training now emphasizes robust protocols for crisis communication, ensuring that all available information is shared and acted upon swiftly.

2. **Decision-Making Under Stress**

   The crew's decisions during the brief emergency were hindered by the overwhelming nature of the fire and the lack of specific training for such scenarios. Modern CRM practices include decision-making frameworks that help crews prioritize actions, even in highly dynamic and time-constrained situations.

Industry Reforms and Lessons Learned

The ValuJet crash prompted significant changes in aviation safety, particularly concerning hazardous materials, airline oversight, and emergency preparedness. Key reforms include:

1. **Stricter Regulations on Hazardous Materials**

   Following the crash, the FAA introduced strict guidelines for transporting hazardous materials, including enhanced labeling, packaging, and documentation requirements. Airlines were also required to implement rigorous checks to ensure compliance.

2. **Enhanced Oversight of Low-Cost Carriers**

   The crash highlighted the risks associated with cost-cutting practices in the aviation industry. The FAA increased its oversight of low-cost carriers, focusing on maintenance practices, contractor accountability, and operational safety standards.

3. **Improvements in Cargo Safety**

   The crash spurred advancements in fire suppression systems and smoke detection technologies for cargo holds. Modern aircraft have fire-resistant materials and more robust containment measures to prevent similar disasters.

4. **Expanded CRM Training**

   CRM training was further refined to address crisis management and emergency decision-making. Flight crews are now trained to handle unexpected scenarios, such as in-flight fires, through simulations emphasizing teamwork, clear communication, and decisive action.

Legacy of ValuJet Flight 592

The ValuJet crash remains a sobering reminder of the catastrophic consequences of systemic failures in aviation safety. It serves as a case study for how lapses in maintenance, oversight, and crisis management can converge to create a tragedy. While the 110 lost lives cannot be recovered, the lessons learned from the disaster have led to industry-wide improvements that continue to save lives today.

By integrating CRM principles, strengthening regulatory frameworks, and enhancing emergency preparedness, the aviation industry has taken meaningful steps to honor the memory of those who perished on ValuJet Flight 592. This tragedy underscores the collective responsibility of airlines, regulators, and crews to prioritize safety above all else.

United Airlines Flight 173 (1978)

On December 28, 1978, United Airlines Flight 173, a McDonnell Douglas DC-8, ran out of fuel and crashed near Portland, Oregon, killing 10 of the 189 passengers and crew onboard. While the aircraft's systems were functioning normally, the crash resulted from a breakdown in communication and decision-making within the cockpit. This tragedy became a landmark case in the development of Crew Resource Management (CRM) and highlighted the importance of addressing human factors in aviation safety.

The Incident

United Airlines Flight 173 was a scheduled flight from Denver, Colorado, to Portland, Oregon, with a routine stop in Portland. As the aircraft approached Portland International Airport, the crew encountered an issue with the landing gear. The gear deployed successfully, but a malfunction in the indicator system left the crew uncertain about whether the gear was securely locked in place.

The captain, a highly experienced pilot with over 27,000 flight hours, chose to circle the airport to troubleshoot the issue, consulting the flight crew and air traffic control. During this time, the aircraft's fuel reserves were critically low. Despite repeated reminders from the flight engineer about the diminishing fuel supply, the captain remained focused on diagnosing the landing gear problem.

After circling for over an hour, the engines began to flame out as the fuel was exhausted. The aircraft crashed into a wooded area approximately six miles from the airport, breaking apart on impact. While most passengers survived, 10 lives were lost, and dozens were injured.

Contributing Factors

1. **Poor Communication and Hierarchical Dynamics**

Investigators found that the cockpit crew's communication was ineffective. The flight engineer repeatedly alerted the captain to the critical fuel situation, but these warnings were either ignored or deprioritized. The captain's authority and focus on the landing gear issue prevented a collaborative discussion about alternative actions, such as an immediate landing. This dynamic, known as "authority gradient," discouraged the first officer and flight engineer from assertively challenging the captain's decisions.

2. **Lack of Situational Awareness**

   The captain's fixation on the landing gear issue led to a loss of situational awareness. He failed to recognize that the fuel situation was equally, if not more, critical. This single-minded focus on one problem—often referred to as "tunnel vision"—is a common human factor in aviation accidents.

3. **Decision-Making Under Stress**

   The crew's decision-making process broke down under the pressure of the landing gear issue and time constraints. The absence of a systematic approach to prioritizing problems and considering alternative actions contributed to the crash.

4. **Inadequate CRM Practices**

   At the time of the accident, formal CRM training was not yet widely implemented in the aviation industry. The cockpit culture did not encourage open dialogue or assertive communication among crew members, limiting the effectiveness of the team as a whole.

Lessons Learned

The crash of United Airlines Flight 173 was a watershed moment for the aviation industry, underscoring the need to address human factors and cockpit dynamics. Key lessons include:

1. **The Importance of Prioritizing Safety Threats**

   The incident highlighted the need for crews to prioritize safety-critical issues, such as fuel levels, over less immediate concerns. CRM training now includes threat and error management frameworks, helping crews identify and address the most urgent risks in real-time.

2. **Empowering All Crew Members**

   The crash demonstrated the dangers of excessive difference to authority in the cockpit. CRM emphasizes flattening the authority gradient, encouraging all crew members to voice concerns assertively, and ensuring that captains foster a collaborative environment.

3. **Systematic Decision-Making**

   Modern CRM training incorporates structured decision-making models, such as the DECIDE model (Detect, Estimate, Choose, Identify, Do, Evaluate), to help crews manage complex scenarios effectively. These models ensure that all available information is considered and actions are evaluated for their impact on overall safety.

4. **Improved Monitoring and Cross-Checking**

   The flight engineer's repeated warnings about the fuel situation went unheeded due to a lack of structured cross-checking protocols. CRM emphasizes the role of monitoring and cross-checking to ensure that critical information is communicated, verified, and acted upon.

Impact on the Aviation Industry

United Airlines Flight 173 directly influenced the development and implementation of CRM training. In the aftermath of the crash, the National Transportation Safety Board (NTSB) recommended that airlines adopt training programs to improve crew communication, teamwork, and decision-making. These recommendations laid the foundation for modern CRM practices, which have become a cornerstone of aviation safety worldwide.

Key reforms and advancements inspired by this tragedy include:

1. **Mandatory CRM Training**

   Following the accident, the FAA required airlines to implement CRM training programs for all flight crews. These programs emphasize open communication, assertiveness, and mutual respect among team members, regardless of rank or experience.

2. **Advancements in Flight Simulator Training**

   Airlines began incorporating realistic, scenario-based training into flight simulators, allowing crews to practice managing complex emergencies in a safe and controlled environment. These scenarios often include hierarchical dynamics, tunnel vision, and conflicting priorities.

3. **Inclusion of Human Factors in Safety Protocols**

   The NTSB and FAA emphasized human factors in aviation safety, leading to the development of policies and procedures designed to mitigate the impact of cognitive and interpersonal challenges in the cockpit.

Broader Lessons for High-Stakes Industries

The lessons from United Airlines Flight 173 extend beyond aviation. CRM principles have been adapted in other high-stakes fields, including healthcare, maritime operations, and emergency services. For example, in healthcare, CRM-inspired communication protocols, such as checklists and team briefings, have reduced surgical errors and improved patient outcomes.

Legacy of United Airlines Flight 173

While the crash of United Airlines Flight 173 was a tragic loss, its impact on aviation safety has been profound. The lessons learned from this incident have saved countless lives by fostering a culture of collaboration, accountability, and continuous improvement. The event is a stark reminder of the importance of addressing human factors and ensuring that all voices are heard in critical decision-making processes.

By embedding CRM principles into every aspect of flight operations, the aviation industry continues to honor the memory of those who perished and ensures that such a tragedy is not repeated.

Military Origins and Broader Applications of CRM

CRM was inspired by military debriefing practices, which emphasize structured analysis and team dynamics to identify areas for improvement. In aviation, the systematic application of these principles has drastically reduced incidents related to human error. Over time, CRM principles have been adopted in other high-stakes industries such as healthcare, maritime operations, and emergency services. For instance, healthcare professionals now use CRM-inspired communication checklists to prevent surgical errors, and maritime crews use similar frameworks to enhance coordination during complex operations.

The cultural shift introduced by CRM has profoundly impacted aviation safety. Today, CRM training is mandated worldwide, ensuring that lessons from tragedies like Tenerife, ValuJet, and United Airlines Flight 173 continue to inform and strengthen safety protocols.

The evolution of CRM represents one of the most significant advancements in aviation safety, directly addressing human factors and transforming cockpit culture. CRM has saved countless lives by fostering communication, collaboration, and critical thinking and serves as a model for other industries. While devastating, the tragedies that led to its development have left a legacy of resilience and continuous improvement.

Simulation in Aviation: A Revolution in Competency and Mastery

The aviation industry also led the way in leveraging simulation for competency and mastery. Flight simulators, introduced in the 1920s with devices like the Link Trainer, evolved into highly sophisticated tools capable of replicating complex flight scenarios. By the 1970s, simulators incorporated advanced technologies that could mimic weather conditions, mechanical failures, and emergency scenarios. These developments paralleled military efforts to use simulation for training combat pilots, further strengthening the ties between military and civilian practices.

Flight simulation allowed pilots to practice high-stakes scenarios in a controlled, risk-free environment, ensuring they mastered critical skills before taking command of actual aircraft. For example, simulators trained pilots to handle engine failures, navigate severe turbulence, and execute emergency landings. This realistic, hands-on training enhanced technical expertise and decision-making under pressure—skills that aligned with the principles of CRM.

The Integration of CRM and Simulation

Integrating CRM and simulation created a powerful framework for addressing technical and human factors in aviation. Structured debriefings became a routine component of simulator-based training sessions, enabling crews to reflect on their performance, identify strengths and weaknesses, and develop actionable strategies for improvement. These debriefings were guided by CRM principles, ensuring that all crew members, from captains to first officers, participated in discussions as equals.

Simulation also reinforced psychological safety, a key element of CRM. By practicing in simulated environments, crews could experiment with communication strategies, practice assertiveness, and learn to navigate hierarchical dynamics without fearing real-world consequences. For instance, copilots were encouraged to challenge captains' decisions during simulated emergencies, fostering a culture of mutual accountability and respect.

Impact on Aviation Safety and Global Influence

The widespread adoption of CRM and simulation transformed aviation into one of the safest modes of transportation. Studies have shown that CRM-based training significantly reduced errors related to communication breakdowns and poor decision-making, contributing to a dramatic decline in aviation accidents over the past few decades (Helmreich & Merritt, 1998). The success of these practices also inspired their adoption in other high-stakes industries, including healthcare, where simulation and debriefing now play a critical role in enhancing patient safety and team dynamics.

Example of a civilian debrief
Photograph courtesy of Will Ratliff

Influence on Other High-Stakes Fields

The success of military-inspired debriefing practices in aviation paved the way for their adoption in other high-stakes civilian fields, most notably healthcare. Like aviation, healthcare is a domain where teamwork, communication, and rapid decision-making are critical to success. The parallels between the two industries made military-derived techniques particularly well-suited to healthcare environments.

Healthcare Simulation

Healthcare simulation emerged as a powerful tool for training medical professionals, and debriefing quickly became a cornerstone of simulation-based education. Drawing directly from military practices, healthcare debriefings emphasize structured reflection, feedback, and team-based learning. For instance, after a simulated cardiac arrest scenario, participants engage in debriefings to analyze their actions, identify communication gaps, and refine clinical skills. This practice aligns closely with the AAR framework, emphasizing a process-oriented approach to learning (Fanning & Gaba, 2007).

One illustrative example comes from the implementation of debriefing protocols in obstetrics. Teams practicing emergency cesarean deliveries in high-fidelity simulations often engage in debriefings to evaluate communication during critical moments, such as the decision to transition from natural delivery to surgery. These discussions reveal insights into timing, role clarity, and the flow of information, leading to improvements in real-world outcomes. Hospitals that adopted such debriefing techniques reported significant reductions in complications and improved teamwork among multidisciplinary teams.

Emergency Response

Another civilian field profoundly influenced by military debriefing practices is emergency response. Firefighters, paramedics, and police officers operate in high-stakes environments where split-second decisions can save lives. Incorporating structured debriefings into their workflows has improved individual performance and team dynamics.

For example, critical incident stress debriefings (CISDs), adapted from military psychological debriefings, have been widely adopted by emergency response teams. These sessions allow first responders to process traumatic experiences, share insights, and support one another. A paramedic recounting their role in a mass casualty event might describe the emotional toll of triaging victims, while their peers offer validation and share similar experiences. This approach not only enhances operational learning but also promotes emotional resilience, reducing the risk of burnout and post-traumatic stress disorder (PTSD).

Case Study: The Implementation of AARs in Firefighting

After Action Reviews have proven particularly effective in the firefighting community, after major incidents, such as large-scale wildfires, teams conduct AARs to evaluate their strategies and tactics. These sessions include both operational and logistical analyses, such as assessing resource allocation and communication effectiveness. One notable success story comes from California's wildfire response teams, which used AARs to identify critical gaps in coordination between state and federal agencies. The lessons learned from these debriefings led to implementing new protocols, significantly improving response times and resource management in subsequent fire seasons.

Firefighters in these AARs often recount harrowing moments, such as being trapped by shifting winds or facing equipment failures in critical situations. By candidly discussing these experiences, teams refine their operational strategies and build trust and camaraderie, reinforcing their collective resilience in the face of future challenges.

Key Takeaways

- Military debriefing practices, particularly structured feedback and reflection have profoundly influenced civilian sectors such as aviation, healthcare, and emergency response.
- The development of Crew Resource Management in aviation demonstrated the effectiveness of military-inspired techniques in improving teamwork, communication, and safety.
- High-stakes civilian fields have successfully adapted debriefing practices to address operational, psychological, and teamwork challenges.
- Case studies, such as AARs in firefighting, illustrate the transformative impact of structured debriefing on performance and coordination.

Reflection

Transitioning military debriefing practices to civilian sectors highlights the universal value of structured feedback and collaborative learning. Initially developed for the battlefield, these principles have found new relevance in saving lives and improving outcomes across diverse industries. Reflecting on this evolution invites us to consider how other military practices might similarly benefit civilian contexts. It also underscores the importance of adapting proven techniques to meet the unique demands of each field.

The stories of pilots navigating crises, healthcare teams refining life-saving procedures, and firefighters braving infernos all demonstrate the power of debriefing to transform lessons into action. As the book continues, we will explore the broader implications of these practices and their potential to drive innovation and resilience in both military and civilian settings.

References

Aviation Safety Network. (n.d.). Accident description: Tenerife disaster. Retrieved from https://aviation-safety.net

Civil Aviation Authority. (2018). Crew resource management (CAP 737). Retrieved from https://publicapps.caa.co.uk

Degani, A., & Wiener, E. L. (1994). *Philosophy, policies, procedures, and practices: The four Ps of flight deck operations.* Human Factors, 36(3), 398–414.

Fanning, R. M., & Gaba, D. M. (2007). The role of debriefing in simulation-based learning. *Simulation in Healthcare, 2*(2), 115-125.

Federal Aviation Administration (FAA). (1997). *Hazardous Materials Regulations and Enforcement after ValuJet Flight 592.* Retrieved from FAA.gov

Helmreich, R. L., & Merritt, A. C. (1998). *Culture at Work in Aviation and Medicine: National, Organizational, and Professional Influences.* Ashgate Publishing.

Helmreich, R. L., Merritt, A. C., & Wilhelm, J. A. (1999). *The evolution of Crew Resource Management training in commercial aviation.* International Journal of Aviation Psychology, 9(1), 19-32.

National Transportation Safety Board (NTSB). (1997). *Aircraft Accident Report: ValuJet Airlines Flight 592 McDonnell Douglas DC-9-32, N904VJ. In-flight fire and impact with terrain, Everglades, near Miami, Florida, May 11, 1996.* NTSB/AAR-97/06. Retrieved from NTSB.gov

National Transportation Safety Board (NTSB). (1979). *Aircraft Accident Report: United Airlines Flight 173 McDonnell Douglas DC-8-61, Portland, Oregon, December 28, 1978.* NTSB/AAR-79/07. Retrieved from NTSB.gov

National Transportation Safety Board (NTSB). (1978). Lessons learned from the Tenerife airport disaster. Retrieved from https://www.ntsb.gov

Wiener, E. L., & Kanki, B. G. (1993). *Cockpit Resource Management.* Academic Press.

Chapter 5
Early Use of Simulation and Debriefing in Healthcare

The Rise of Simulation-Based Training in Healthcare

The integration of simulation into healthcare education marked a transformative shift in how medical professionals were trained. Historically, healthcare training relied heavily on a traditional apprenticeship model, where students learn by observing and participating in real patient care under supervision. While effective in some respects, this approach offered limited opportunities for hands-on practice in a safe and controlled environment. Simulation-based training emerged as a solution to bridge this gap, providing realistic, risk-free settings where learners could develop and refine their skills.

The use of medical simulators dates to the mid-20th century. One of the earliest and most notable examples is Resusci Anne, a mannequin introduced in the 1960s for cardiopulmonary resuscitation (CPR) training. This innovation allowed healthcare providers to practice life-saving techniques repeatedly, mastering the skills needed to respond effectively in emergencies. Stories of trainees overcoming initial hesitations and mastering CPR on Resusci Anne illustrate the mannequin's profound impact on preparedness and confidence. Over the decades, simulation technology has advanced significantly, incorporating high-fidelity mannequins, virtual reality, and immersive scenarios. These tools enabled learners to simulate complex medical procedures, team-based interventions, and critical decision-making processes.

By the late 20th century, simulation centers emerged as integral components of medical education. Institutions like the Center for Medical Simulation (CMS) in Boston and Stanford University's simulation program pioneered the development and application of simulation-based methodologies. At these centers, medical students practiced high-stakes scenarios—from handling a code blue to managing mass casualty events—without risking patient safety. Participants often described the experience as "stressful but invaluable," with simulation mistakes becoming learning opportunities rather than threats to patient well-being. These centers demonstrated that simulation could enhance technical proficiency and improve communication, teamwork, and patient safety. The rise of simulation-based training laid the groundwork for integrating debriefing as a core educational practice in healthcare.

Challenges in Translating Military Debriefing to Medical Settings

Adapting military-style debriefing practices to healthcare was not without its challenges. While the principles of structured feedback, reflection, and continuous improvement were universally applicable, the context and culture of healthcare required thoughtful modifications to these practices.

One significant challenge was the hierarchical nature of medical teams. In the military, rank, and structure are integral to operations, but military debriefing practices emphasize collaboration and equal input from all participants. In healthcare, deeply ingrained hierarchies often made it difficult for junior team members, such as medical students or nurses, to speak up during debriefings. For instance, in early implementations of debriefing, junior staff members were hesitant to point out communication failures by senior surgeons. This dynamic risked stifling valuable feedback and limiting the effectiveness of the learning process.

Another challenge was the emotional intensity of healthcare work. Unlike military operations, which are typically conducted away from civilian populations, healthcare professionals often work directly with patients and their families. This proximity to human suffering added an emotional dimension to debriefings that required careful navigation. Facilitators had to balance the need for critical feedback with the importance of psychological safety, ensuring that participants felt supported while reflecting on emotionally charged experiences. In one case, a debrief following a failed resuscitation required the facilitator to address not only procedural missteps but also the visible grief among team members.

Lastly, the diverse educational backgrounds of healthcare teams posed a challenge. Unlike military units, which undergo standardized training, healthcare teams often consist of individuals with varying levels of expertise and different professional frameworks. For example, a nurse trained in rapid response protocols might have a different approach than a physician focused on diagnosis. Bridging these gaps required facilitators to tailor debriefing sessions to meet the needs of diverse participants, fostering inclusivity and mutual understanding. Over time, these challenges were met with innovative solutions, such as pre-briefing participants to set expectations and using debriefing scripts to guide discussions.

Challenge	Description	Solutions
Hierarchical nature of medical teams	Ingrained hierarchies often make it difficult for junior team members to provide feedback, limiting the effectiveness of debriefings.	Pre-briefing participants to set expectations; fostering a culture of open dialogue.
Emotional intensity of healthcare work	Healthcare work involves direct interaction with patients and families, adding emotional complexity to debriefings.	Ensuring psychological safety during debriefings; addressing emotional responses with sensitivity.
Diverse educational backgrounds	Healthcare teams have varying levels of expertise and training, requiring tailored approaches to ensure inclusivity and mutual understanding.	Using debriefing scripts to guide discussions; tailoring sessions to the needs of diverse participants.

Pioneering Figures and Institutions in Healthcare Simulation

The rise of simulation and debriefing in healthcare can be traced to the vision and dedication of a few trailblazers who saw beyond the traditional methods of medical education. Their belief in the transformative potential of simulation-based learning and their relentless efforts to promote its adoption laid the foundation for what is now a cornerstone of modern healthcare training.

One of the most influential figures in this movement was Dr. David Gaba, often hailed as the "father of simulation-based healthcare education." In the 1980s, Dr. Gaba was an anesthesiologist grappling with the realization that rare but life-threatening scenarios—such as malignant hyperthermia or sudden equipment failure—were not adequately addressed in traditional training. He envisioned a way to prepare anesthesia providers for these critical moments without risking patient safety. This vision led to the development of the Comprehensive Anesthesia Simulation Environment (CASE), a groundbreaking program that allowed healthcare professionals to practice responding to high-stakes situations in a controlled environment.

Dr. Gaba's work did not stop recreating scenarios. He recognized that the real learning happened during the debriefing process after the simulation. This reflective practice allowed participants to analyze their actions, discuss what went right and wrong, and identify ways to improve. His approach revolutionized medical education, shifting the focus from merely acquiring technical skills to understanding the human factors—like communication, decision-making, and teamwork—that are often the difference between life and death in a clinical setting.

While Dr. Gaba pioneered simulation in anesthesia, another visionary, Dr. Jeffrey Cooper, delved into how human factors impacted patient safety. Dr. Cooper's focus was broader, examining how teamwork, leadership, and communication influenced outcomes across healthcare disciplines. His work laid the groundwork for applying simulation beyond individual specialties, addressing the systemic challenges of healthcare delivery.

These ideas took root in institutions dedicated to simulation-based education. The Center for Medical Simulation (CMS) in Boston, founded in 1993, became one of the first centers to bring these concepts to life. Within its walls, healthcare teams participated in high-fidelity simulations that mirrored real-life crises—such as cardiac arrests or neonatal emergencies. After each scenario, the teams would sit down for debriefing sessions led by skilled facilitators. These intense, eye-opening discussions helped participants uncover blind spots in their practice and understand how their actions and communication impacted outcomes.

The Mayo Clinic, another institution at the forefront of simulation, took a different approach by emphasizing interprofessional training. Its Multidisciplinary Simulation Center, established in the early 2000s, brought together diverse healthcare providers—physicians, nurses, and respiratory therapists—to work through complex scenarios as a team. In one memorable simulation, a deteriorating patient required the team to coordinate seamlessly under time pressure. The subsequent debriefing revealed the critical role of clear communication and role clarity in achieving positive outcomes. At Mayo, the focus was on individual performance and how teams functioned as a unit, reflecting the real-world dynamics of healthcare delivery.

As simulation gained traction, professional organizations emerged to formalize and expand its impact. The Society for Simulation in Healthcare (SSH), founded in 2004, became a rallying point for educators, researchers, and practitioners. SSH brought simulation into the mainstream through conferences, standards, advocacy, showcasing cutting-edge technologies and best practices. Meanwhile, organizations like the International Nursing Association for Clinical Simulation and Learning (INACSL) and the Society for Academic Emergency Medicine (SAEM) pushed the boundaries of what simulation could achieve in specific fields, from nursing to emergency medicine.

These pioneers and institutions didn't just innovate; they transformed. By making simulation a platform for reflection, collaboration, and continuous improvement, they reshaped how healthcare professionals prepare for the challenges of modern medicine. Today, their legacy lives on in every simulated crisis, every debriefing discussion, and every healthcare professional who steps into a real-world emergency better prepared to save lives.

Influential Leaders

Dr. David Gaba

Contributions:
Often regarded as the "father of simulation-based healthcare education," Dr. Gaba pioneered the use of simulation and debriefing in clinical training. In the 1980s, he developed the **Comprehensive Anesthesia Simulation Environment (CASE)**, one of the first high-fidelity simulation systems for training anesthesia providers. His research emphasized the integration of debriefing into simulation exercises to enhance reflective learning and improve patient safety (Gaba, 2004).

Impact:
Dr. Gaba's work laid the foundation for modern healthcare simulation centers, and his advocacy for simulation-based teamwork training influenced practices across specialties.

Jenny Rudolph, PhD

Contributions:
Dr. Rudolph is a leader in healthcare simulation and debriefing research. As Executive Director of the Center for Medical Simulation (CMS) in Boston, she co-developed the **Debriefing with Good Judgment** model, which combines advocacy and inquiry techniques to facilitate reflective learning. This approach addresses emotional and cognitive barriers to effective debriefing, helping participants critically analyze their performance while maintaining psychological safety.

Impact:
Her work has been instrumental in teaching healthcare educators how to conduct meaningful, nonjudgmental debriefings that foster growth and collaboration.

Dr. Walter Eppich

Contributions:
A pediatric emergency medicine physician and simulation educator, Dr. Eppich is renowned for his contributions to team-based learning and

debriefing in healthcare. He has extensively studied **debriefing frameworks**, including the application of conversational analysis to improve feedback and reflection during debriefings. Dr. Eppich's research highlights the importance of adaptive debriefing strategies tailored to diverse team dynamics.

Impact:
Dr. Eppich's work has informed the development of evidence-based debriefing practices, particularly in high-stakes, interprofessional scenarios.

Dr. Adam Cheng

Contributions:
Dr. Cheng is a prominent figure in pediatric simulation and debriefing. He co-led the **International Network for Simulation-based Pediatric Innovation, Research, and Education (INSPIRE)** and conducted groundbreaking research on the effectiveness of debriefing methodologies. His contributions include the development of tools to assess debriefing quality, such as the **Debriefing Assessment for Simulation in Healthcare (DASH)**.

Impact:
Dr. Cheng's work has provided healthcare educators with standardized approaches to evaluate and enhance debriefing effectiveness.

Ruth D. Wilson, MD

Contributions:
Dr. Wilson has been a leading advocate for integrating simulation and debriefing into obstetrics and gynecology. She developed **team-based simulation programs** to improve outcomes in maternal and neonatal care, emphasizing the role of debriefing in addressing adverse events and enhancing interprofessional collaboration.

Impact:
Her initiatives have improved obstetric crisis management and fostered a culture of continuous learning in maternal health settings.

Dr. Vincent Grant

Contributions:
As a pediatric intensivist and simulation leader, Dr. Grant has focused on improving the design and execution of debriefing in healthcare education. He has contributed to the development of **structured debriefing models** and advocated for using simulation as a quality improvement tool.

Impact:
Dr. Grant's research and teaching have influenced healthcare systems worldwide, integrating debriefing into patient safety and crisis management practices.

Dr. Elizabeth Hunt

Contributions:
Dr. Hunt, a pediatric critical care specialist, has conducted pioneering work in simulation-based education and crisis resource management. She developed simulation curricula for resuscitation teams, emphasizing the integration of **psychological safety** and structured debriefing to improve teamwork and outcomes in emergency settings.

Impact:
Her efforts have advanced the understanding of how debriefing can improve both technical and non-technical skills in critical care environments.

Dr. Jeffrey Cooper

Contributions:
A biomedical engineer by training, Dr. Cooper is a pioneer in patient safety and simulation. He was instrumental in founding the **Center for Medical Simulation (CMS)** in Boston, one of the first simulation centers dedicated to healthcare education. His work emphasized the critical role of debriefing in learning from simulation exercises and real-world events, particularly in the context of anesthesia safety.

Impact:
Dr. Cooper's leadership helped establish simulation as a core component of patient safety initiatives and inspired the development of structured debriefing practices.

Dr. Mary Fey

Contributions:
Dr. Fey is a recognized leader in simulation-based education and debriefing, particularly for nursing. She has researched and developed frameworks for effective debriefing, focusing on psychological safety and learner-centered approaches. Her contributions to the **Debriefing Assessment for Simulation in Healthcare (DASH)** tool have helped standardize the evaluation of debriefing quality.

Impact:
Dr. Fey's work has elevated the role of simulation in nursing education and supported healthcare educators in delivering high-quality debriefings.

Dr. Pamela Andreatta

Contributions:
Dr. Andreatta has been a leading voice in using simulation and debriefing for surgical training and systems improvement. Her research focuses on the integration of **high-fidelity simulation** into healthcare workflows to enhance both technical and non-technical skills. She has developed simulation-based curricula for various specialties, incorporating structured debriefing to optimize learning outcomes.

Impact:
Her contributions have advanced the application of simulation and debriefing in surgical education, improving procedural competence and team performance.

Dr. Robert Simon

Contributions:
Dr. Simon co-developed the **Debriefing with Good Judgment** framework alongside Dr. Jenny Rudolph. He has also been a key figure at the Center for Medical Simulation (CMS), where he has trained healthcare educators to facilitate meaningful, psychologically safe debriefings.

Impact:
Dr. Simon's work has emphasized the importance of balancing advocacy with inquiry in debriefing, empowering learners to reflect on their performance critically.

Dr. Elizabeth Armstrong

Contributions:
Dr. Armstrong has been an influential advocate for simulation and debriefing in medical education, particularly through her leadership at Harvard Medical School. She has developed simulation-based teaching programs that incorporate structured debriefing to enhance clinical reasoning and interprofessional collaboration.

Impact:
Her efforts have helped establish simulation and debriefing as integral components of curriculum design at leading academic institutions.

Dr. Debra Nestel

Contributions:
Dr. Nestel is a global leader in healthcare simulation, based in Australia. She has developed numerous simulation programs that integrate debriefing for interprofessional education and patient safety. Her work includes publishing widely on simulation and debriefing practices, including the development of innovative debriefing strategies for diverse healthcare settings.

Impact:
Dr. Nestel's contributions have had a global influence, particularly in advancing debriefing practices in resource-constrained environments.

Dr. George A. Miller

Contributions:
Dr. Miller, known for his foundational work in medical education, contributed to the development of simulation as a tool for competency-based assessment. His "Miller's Pyramid" model, which outlines stages of clinical competence (from "knows" to "does"), is often used in simulation debriefings to evaluate learner performance and identify areas for growth.

Impact:
Though not exclusively focused on debriefing, Dr. Miller's work has shaped how simulation and feedback assess and enhance clinical competence.

Dr. Rebecca Minehart

Contributions:
Dr. Minehart is an expert in simulation and debriefing in obstetric anesthesia. She has researched using simulation and debriefing to improve team communication, crisis resource management, and maternal safety. Her work focuses on integrating reflective learning into practice, particularly in high-stakes environments.

Impact:
Her contributions have improved team dynamics and patient safety in obstetric care through simulation-based education.

Dr. Amitai Ziv

Contributions:
Dr. Ziv founded the Israel Center for Medical Simulation (MSR) and is a leading advocate for simulation-based learning and debriefing in healthcare. His initiatives focus on using simulation to address human error and promote a culture of safety.

Impact:
Dr. Ziv's work has made simulation and debriefing central to medical training in Israel and inspired similar programs worldwide.

Dr. John Schaefer

Contributions:
Dr. Schaefer has been a leader in anesthesia simulation and the development of simulation centers. He co-developed advanced simulation models for crisis management and contributed to training curricula integrating structured debriefing.

Impact:
His work has advanced simulation-based education in anesthesiology and critical care, promoting debriefing as a tool for continuous improvement.

Example of a healthcare simulation. Note they are using video to debrief.
Photograph courtesy of Global Network for Simulation in Healthcare

Key Takeaways

1. The rise of simulation-based training in healthcare provided a safe, controlled environment for medical professionals to practice and

refine their skills, addressing the limitations of traditional training models.

2. Translating military debriefing practices to healthcare requires overcoming challenges related to hierarchy, emotional intensity, and diverse educational backgrounds.
3. Pioneering figures like Dr. David Gaba and institutions like the Center for Medical Simulation played critical roles in advancing simulation and debriefing in healthcare education.
4. Establishing dedicated simulation centers and organizations like the Society for Simulation in Healthcare (SSH) laid the foundation for the widespread adoption of these practices.

Reflection

The early adoption of simulation and debriefing in healthcare highlights the transformative power of these tools in improving education and patient care. Reflecting on this journey, we see the importance of innovation, collaboration, and adaptability in addressing complex challenges. The work of pioneers and institutions revolutionized medical training and demonstrated the universal value of structured reflection and feedback. As we continue to explore the intersection of simulation and debriefing, we are reminded of the enduring impact of these practices on healthcare professionals and the patients they serve.

References

Gaba, D. M. (2004). The future vision of simulation in health care. *Quality and Safety in Health Care, 13*(suppl 1), i2–i10.

Chapter 6 The Role of Crisis in Accelerating Change

How Medical Errors and Crises Drove the Adoption of Debriefing

Throughout history, crises have often acted as powerful catalysts for change, forcing industries and institutions to confront their vulnerabilities and seek solutions. For healthcare, recognizing medical errors as a systemic issue became one of the most profound drivers of change. The pivotal moment in this realization came with the publication of the Institute of Medicine's (IOM) groundbreaking report, *To Err is Human: Building a Safer Health System*, in 1999.

The report was a revelation. It estimated that medical errors accounted for up to 98,000 deaths annually in the United States, placing them among the leading causes of mortality. These errors were not due to incompetence or lack of knowledge but were often systemic failures—miscommunication, inadequate protocols, poor teamwork, and a lack of accountability. The findings were alarming and galvanizing for healthcare professionals and the public alike. The report became a rallying cry for change, underscoring the urgent need to adopt practices that could address these issues head-on.

Among the solutions proposed, the concept of debriefing stood out. Already well-established in aviation and military operations, where lives depended on clear communication and constant improvement, debriefing offered a structured way to reflect on events, analyze mistakes, and foster a culture of accountability. The principles of debriefing—open dialogue, collaborative problem-solving, and a focus on learning rather than blame—aligned perfectly with the challenges highlighted by *To Err is Human*.

The Spread of Debriefing in Healthcare

In the years following the report, healthcare institutions began incorporating debriefing into their processes to address errors and improve safety. Hospitals started conducting post-incident reviews, where teams could come together to analyze adverse events and identify their root causes. These debriefing sessions were designed to break down silos and encourage honest, constructive conversations among multidisciplinary teams.

One of the most significant areas of focus was miscommunication. For example, handoffs between providers—when critical information about a patient's condition is transferred—were identified as a frequent source of errors. In debriefings, teams began to dissect these moments, identifying what went wrong and developing strategies to improve. Simple but impactful changes, such as using standardized handoff tools like the SBAR (Situation, Background, Assessment, Recommendation) framework, emerged directly from these reflections.

Real-Life Examples of Change

The transformative potential of debriefing is perhaps best illustrated through specific examples. At one pediatric hospital, a devastating medication error nearly claimed the life of a young patient. A nurse had administered a dosage ten times higher than prescribed due to miscommunication with the pharmacy and an unclear labeling system. The near miss triggered a post-incident debrief involving nurses, pharmacists, and administrators.

The team examined the events leading to the error in the debriefing session. They discovered that multiple points of failure had converged: the pharmacy's labeling practices were inconsistent, the nurse's workload had led to cognitive overload, and there was no standardized double-check protocol for high-risk medications. The discussion was candid but constructive, focusing on system improvements rather than individual blame.

As a result, the hospital implemented a series of changes. The pharmacy updated its labeling system to include clearer warnings for high-risk medications. Nurses were trained to use a two-person verification system before administering certain drugs. Automated alerts were added to the hospital's electronic medical records to flag potential errors in dosage calculations. Over time, the hospital saw a dramatic reduction in medication errors, a direct result of the changes prompted by the debrief.

Cultural Shifts and the Role of Debriefing

Beyond addressing specific incidents, debriefing began to influence the broader healthcare culture. Traditionally, the medical field operated within rigid hierarchies, where questioning authority or admitting mistakes was often discouraged. This culture contributed to the very problems highlighted in To Err is Human. Debriefing introduced a new paradigm: one where every team member's voice mattered, mistakes were seen as opportunities to learn, and accountability was shared.

For instance, in operating rooms, surgeons, nurses, anesthesiologists, and technicians began to conduct "hot routine debriefs" immediately after surgeries, particularly in cases where complications occurred. These sessions allowed teams to reflect on what had gone well and what could be improved while the experience was fresh. Over time, this practice fostered trust and respect among team members, breaking down traditional hierarchies and creating a safer, more collaborative environment.

The Broader Impact

As debriefing became more widespread, its impact extended beyond addressing errors. It became a cornerstone of continuous improvement in healthcare, shaping how professionals approached everything from individual patient care to large-scale systemic changes. Multidisciplinary debriefs provided a forum for innovation, enabling teams to identify inefficiencies and develop creative solutions.

Debriefing also began to influence how healthcare providers prepared for crises. Simulation training, which often incorporates debriefing, became a standard tool for teaching teams to manage high-stakes scenarios like cardiac arrests or mass casualty events. These simulations improved technical skills and emphasized communication, teamwork, and decision-making under pressure—critical elements for reducing errors and improving outcomes in real-world situations.

A Legacy of Learning

The publication of *To Err is Human* was a watershed moment for healthcare, forcing the industry to confront its shortcomings and adopt practices that could drive meaningful change. Debriefing emerged as one of the most powerful tools in this transformation. It addressed systemic issues plaguing healthcare for decades, encouraging reflection, accountability, and collaboration.

Today, the lessons learned from the adoption of debriefing continue to shape the field. Hospitals, clinics, and training programs worldwide have embraced its principles, creating an ongoing learning and improvement culture. While the journey to safer healthcare is far from complete, the integration of debriefing has undoubtedly saved countless lives and set a new standard for how the industry responds to errors and crises.

Critical Incident Stress Debriefing (CISM): Bridging Military and Civilian Applications

Introduction to CISM

Critical Incident Stress Debriefing (CISM) is a structured, peer-supported, and facilitator-led intervention designed to help individuals process and recover from traumatic events. Initially developed for military personnel dealing with combat stress, CISM has since been adapted to civilian contexts, including emergency response, healthcare, and disaster management. Its primary goals are to mitigate the immediate psychological impact of critical incidents, promote emotional recovery, and reduce the risk of long-term psychological issues such as post-traumatic stress disorder (PTSD).

CISM in the Military

Origins and Evolution:

CISM originated as a component of battlefield psychology, where it addressed acute stress reactions among soldiers. Military psychologists recognized that discussing the emotional and operational aspects of traumatic events helped soldiers process their experiences and return to duty.

During the Vietnam War, informal debriefings became integral to managing combat fatigue. These evolved into structured models, such as "After Action Reviews" (AARs), with a psychological component.

Key Features in Military Settings:

- Focuses on both operational performance and emotional well-being.
- Conducted shortly after a traumatic event (e.g., combat, casualties, or mission failures) to provide immediate support.
- Emphasizes confidentiality and hierarchy-neutral participation to foster openness.

Tools Used in Group CISM

Demobilization

Demobilization is a brief, informal intervention conducted immediately after a critical incident, designed to help individuals transition from the operational environment to recovery. It is typically used in large-scale situations like natural disasters or mass casualty events. The primary goal is to reduce immediate stress by providing practical information and psychological support. Demobilization sessions often include two components: an informational briefing and stress education. In the briefing, a facilitator updates participants on the status of the incident and addresses any logistical concerns, such as next steps or shifts. The stress education component introduces participants to common stress responses and offers basic self-care strategies, like hydration and rest. These sessions are conducted at the end of a shift or operational phase and serve as a starting point for more comprehensive interventions if needed. For example, firefighters completing their work on a large wildfire might attend a demobilization session to decompress and gain tools for managing immediate stress.

Defusing

Defusing is a short, informal discussion within 8-12 hours of a critical incident. It is designed to provide immediate emotional support and prepare participants for more in-depth debriefing sessions, if necessary. Defusing allows participants to express their initial thoughts and reactions in a safe environment, clarify misconceptions about the incident, and normalize their stress responses. These sessions typically last 20-45 minutes and follow a simple structure: an introduction, exploration, and reassurance. In the introduction, the facilitator sets expectations for the session, emphasizes confidentiality, and outlines the purpose of the discussion. Participants share their roles and immediate concerns about the event during the exploration phase. The session concludes with reassurance, where facilitators provide stress management tips and remind participants that their reactions are normal. Defusing sessions are ideal for mitigating acute stress and reducing the risk of long-term psychological harm. For instance, after a high-stakes police raid, officers might participate in a defusing session to address immediate concerns and emotional reactions before resuming their duties.

Crisis Management Briefing (CMB)

Crisis Management Briefing (CMB) is a large-group intervention designed for situations where an entire community or organization is affected by a traumatic event, such as a natural disaster or school shooting. This type of debriefing aims to provide timely information, alleviate panic or misinformation, and offer resources to promote resilience. A CMB session often begins with an incident overview, offering updates about the situation and recovery efforts. Facilitators then provide supportive information, such as stress management techniques and contact information for additional mental health resources. Finally, the session concludes with a Q&A segment to address participants' concerns and clarify the next steps. These sessions typically last one to two hours and are conducted during or shortly after the crisis. For example, after a tornado devastates a community, a CMB might bring together residents to provide updates on rescue operations, educate them about stress responses, and connect them with available support services.

The 7-Phase Critical Incident Stress Debriefing (CISD)

The 7-Phase CISD is the hallmark of Critical Incident Stress Management. It is a structured, small-group debriefing designed to help participants process the psychological impact of a traumatic event. This approach is typically used within 24-72 hours of the incident once participants have had time to stabilize emotionally. The session follows seven distinct phases, beginning with an **introduction phase** where facilitators establish confidentiality, explain the session's purpose, and set expectations. Participants share objective details about the event from their perspectives in the **fact phase**. This leads into the **thought phase**, where they discuss their initial impressions and reactions to the incident. The

reaction phase delves deeper into emotional responses, allowing participants to express how the event affected them personally. Next, the **symptom phase** focuses on identifying physical, emotional, or behavioral symptoms they have experienced. The **teaching phase** provides education on normal stress reactions and coping strategies to promote emotional recovery. The session concludes with the **re-entry phase**, where facilitators summarize the discussion, address any lingering questions, and outline available follow-up resources.

For example, healthcare workers involved in a mass casualty event might participate in a CISD session to process their emotions, discuss symptoms like sleep disturbances, and learn effective stress management techniques.

Each of these CISM tools—demobilization, defusing, crisis management briefing, and the 7-phase CISD—serves a specific purpose and is tailored to different contexts and timelines. Together, they form a comprehensive framework for helping individuals and groups process traumatic experiences, mitigate stress, and recover emotionally. These methods are versatile and can be adapted based on the needs of participants and the severity of the incident, making them valuable tools in both military and civilian applications.

Tool	**Duration**	**Timing**	**Focus**	**Ideal Context**
Demobilization	15-30 minutes	Immediately after event	Transition from operations to recovery	Large-scale incidents, such as wildfires or disasters
Defusing	20-45 minutes	8-12 hours post-incident	Emotional ventilation, normalization	First responders after a high-stakes event
Crisis Management Briefing	1-2 hours	During or post-crisis	Information dissemination, stress management	Community-wide crises, disasters, or public events
7-Phase CISD	1-3 hours	24-72 hours post-incident	Emotional processing,	Small groups exposed to shared trauma

Tool	Duration	Timing	Focus	Ideal Context
			coping strategies	

Example Use Case:

A special forces unit conducts a CISM session after losing team members in a high-stakes mission. A trained psychologist facilitates the debriefing, allowing participants to share emotions, discuss coping strategies, and identify support for returning to duty.

CISM in Civilian Applications

Adoption by Emergency Responders:

Firefighters, paramedics, police officers, and healthcare workers often face traumatic incidents, such as mass casualty events or violent encounters. CISM provides a structured framework to help these professionals process their experiences. Sessions typically involve individual and team debriefings, promoting collective and personal resilience.

Integration into Healthcare:

CISM is being used more frequently to support healthcare teams dealing with emotionally intense situations, such as patient deaths or medical errors. It aids in reducing burnout and promotes emotional recovery.

Adaptations for Civilian Use:

Civilian CISM often includes additional resources, such as peer support networks, counseling referrals, and training in resilience-building techniques. Sessions are tailored to diverse teams and emphasize inclusivity and interprofessional collaboration.

Example Use Case:

After a school shooting, first responders engage in a CISM session. Facilitators guide participants in expressing their emotions, validating their experiences, and identifying strategies to manage stress and prevent burnout.

Key Principles of CISM

- **Immediacy:** Conducted soon after the incident to address acute stress reactions.
- **Confidentiality:** Creates a safe space for participants to share openly.
- **Non-Judgmental Atmosphere:** Focuses on emotional recovery rather than assigning blame.
- **Normalization:** Helps participants understand that their reactions are normal responses to abnormal situations.
- **Follow-Up Support:** Ensures access to additional resources, such as counseling or ongoing peer support.

Challenges and Criticisms

- **Potential for Harm:** Critics argue that poorly conducted CISM sessions can retraumatize participants by forcing them to relive distressing events.
- **Mixed Evidence:** Research on CISM's efficacy is mixed, with some studies suggesting limited long-term benefits.
- **Facilitator Training:** Success relies heavily on the skill and experience of the facilitator.

Future Directions

- **Technology Integration:** Virtual reality (VR) and telehealth platforms are being explored as tools to deliver CISM in remote or resource-limited settings.
- **Resilience Training:** Combining CISM with proactive resilience-building programs to prepare individuals for high-stress roles.
- **Cross-Disciplinary Learning:** Adapting military CISM practices to better fit civilian contexts by incorporating psychology, healthcare, and organizational behavior insights.

Case Studies of Debriefing During Public Health Emergencies

The 2003 SARS Outbreak

The 2003 Severe Acute Respiratory Syndrome (SARS) outbreak was a defining moment in global public health. Healthcare systems were unprepared for the virus's rapid spread, leading to widespread fear, resource shortages, and significant strain on frontline workers. In the aftermath, debriefing played a crucial role in understanding the failures and successes of the response.

Hospitals in Toronto, one of the hardest-hit cities, conducted extensive debriefings to evaluate their response efforts. These sessions highlighted critical gaps in preparedness, such as inadequate infection control protocols and communication breakdowns between healthcare providers and public health authorities. For instance, one debrief revealed that delays in disseminating updated guidelines led to inconsistent practices across institutions. Based on these findings, hospitals implemented more robust infection control training and established clearer lines of communication, which proved invaluable in future outbreaks like H1N1 and COVID-19.

Frontline workers who participated in these debriefings often spoke about the emotional toll of battling an unknown and deadly virus. Nurses shared stories of exhaustion and fear, while doctors reflected on the ethical challenges of resource allocation. These narratives, captured during debriefings, not only informed systemic changes but also validated the experiences of healthcare providers, fostering resilience and unity within the profession.

The 2014 Ebola Crisis

The 2014 Ebola epidemic in West Africa exposed the vulnerabilities of global healthcare systems in managing highly infectious diseases. The crisis underscored the importance of simulation and debriefing in preparing healthcare workers to handle such cases in the United States. The transfer of two Ebola patients to Emory University Hospital's specialized biocontainment unit became a case study in effective debriefing.

Before the patients' arrival, the medical team engaged in rigorous simulations to practice donning and doffing personal protective equipment (PPE), managing infectious waste, and coordinating patient care. After each simulation, debriefing sessions allowed team members to reflect on their performance, identify potential risks, and adjust protocols. For example, during one debrief, team members recognized that unclear hand signals while wearing PPE hindered communication, leading to the adoption of a standardized gesture system.

Post-incident debriefings further emphasized the value of these practices. The team comprehensively reviewed their responses after the patients were successfully treated and discharged. This debriefing validated their preparation's effectiveness and generated recommendations for other institutions developing biocontainment capabilities. Team members often cited these debriefings as crucial to their confidence and effectiveness during the crisis, highlighting the interplay between technical readiness and psychological preparedness.

The COVID-19 Pandemic

The COVID-19 pandemic, the most significant global health crisis of the 21st century, brought debriefing practices to the forefront of healthcare operations. From the early days of the pandemic, healthcare workers faced unprecedented challenges, including surges in patient volume, PPE shortages, and rapidly evolving treatment protocols.

In response, many institutions implemented regular debriefings to adapt to the rapidly changing landscape. Emergency departments, for instance, held daily debriefings at the end of each shift to review challenges, share insights, and propose solutions. These sessions revealed critical issues, such as the need for streamlined patient triage systems and better strategies for managing staff burnout. Debriefings also provided a platform for frontline workers to share emotional experiences, fostering a sense of solidarity and resilience during an extraordinarily difficult time.

One notable example is a hospital in New York City that established a multidisciplinary debriefing team during the pandemic's peak. This team facilitated sessions with diverse groups, including physicians, nurses, and support staff, ensuring that all perspectives were considered. The insights gained from these debriefings informed changes to patient flow, PPE allocation, and staffing models, ultimately improving the hospital's ability to manage subsequent surges. Staff members often described these debriefings as lifelines, enabling them to feel heard and supported while navigating the crisis.

Key Takeaways

1. Crises have historically served as catalysts for adopting debriefing practices in healthcare, highlighting the need for structured reflection and feedback to address systemic issues such as

communication gaps, teamwork challenges, and accountability deficits.

2. The *To Err is Human* report was a landmark moment that revealed the scope of medical errors and underscored the importance of learning-oriented interventions like debriefing to foster a culture of safety and improvement.
3. Critical Incident Stress Debriefing (CISM) has effectively bridged military and civilian applications, providing structured support for managing the psychological impact of traumatic events across diverse contexts.
4. Case studies, such as the SARS outbreak, Ebola crisis, and COVID-19 pandemic, emphasize the pivotal role of debriefing in improving operational response, team resilience, and preparedness during public health emergencies.
5. Regular debriefings benefit healthcare teams by uncovering latent issues, enhancing emotional recovery, and driving long-term patient care and system readiness improvements.

Reflection

Crises arising from systemic challenges like medical errors or global emergencies such as pandemics often illuminate vulnerabilities within healthcare systems. These moments compel organizations to embrace change, and debriefing has become vital in this transformation. By fostering a structured environment for reflection, debriefing addresses immediate challenges while cultivating resilience and accountability for the future.

The examples provided in this chapter—ranging from the *To Err is Human* report to the global responses to SARS, Ebola, and COVID-19—demonstrate how debriefing enhances learning and operational effectiveness. These cases remind us that despite immense difficulty, there are opportunities to innovate, grow, and improve.

As healthcare systems prepare for future crises, it is critical to adopt debriefing proactively rather than reactively. By embedding structured reflection into routine practices, organizations can strengthen their ability to adapt and thrive under pressure. This chapter serves as a testament to the transformative power of debriefing, urging leaders to prioritize it as a cornerstone of healthcare resilience and innovation.

References

Cheng, A., Grant, V., Dieckmann, P., Arora, S., Robinson, T., & Eppich, W. (2016). Debriefing for technology-enhanced simulation: A systematic review and meta-analysis. Medical Education, 50(4), 469-478.

Dyregrov, A. (1997). The process in critical incident stress debriefings. Journal of Traumatic Stress, 10(4), 589-605.

Everly, G. S., & Mitchell, J. T. (1999). Critical Incident Stress Management (CISM): Advanced concepts. Ellicott City, MD: Chevron Publishing.

Fanning, R. M., & Gaba, D. M. (2007). The role of debriefing in simulation-based learning. Simulation in Healthcare, 2(2), 115-125.

Halamek, L. P. (2012). Simulation-based training in neonatal resuscitation: An evidence-based approach. Seminars in Perinatology, 36(6), 449-457.

Halpern, J., Maunder, R. G., Schwartz, B., & Gurevich, M. (2012). The critical incident stress debriefing process for emergency responders: Review of effectiveness. International Journal of Emergency Mental Health, 14(4), 253-265.

Institute of Medicine. (1999). To Err is Human: Building a Safer Health System. National Academies Press.

Mitchell, J. T. (1983). When disaster strikes: The critical incident stress debriefing process. Journal of Emergency Medical Services, 8(1), 36-39.

Mitchell, J. T., & Everly, G. S. (2001). Critical Incident Stress Management (CISM): A new era and standard of care in crisis intervention. Ellicott City, MD: Chevron Publishing.

National Institute for Occupational Safety and Health (NIOSH). (2002). Stress... at work. Washington, DC: U.S. Department of Health and Human Services.

Patterson, M. D., Blike, G. T., & Nadkarni, V. M. (2008). In situ simulation: Challenges and results. Simulation in Healthcare, 3(3), 140-146.

Rudolph, J. W., Simon, R., Dufresne, R. L., & Raemer, D. B. (2006). There's no such thing as 'nonjudgmental' debriefing: A theory and method for debriefing with good judgment. Simulation in Healthcare, 1(1), 49-55.

Skeffington, P. M., Rees, C. S., Mazzucchelli, T. G., & Kane, R. T. (2017). The primary prevention of PTSD: A systematic review. Journal of Trauma & Dissociation, 18(1), 88-105.

Tuckey, M. R., & Scott, J. E. (2014). Group critical incident stress debriefing with emergency services personnel: A randomized controlled trial. Anxiety, Stress, & Coping, 27(1), 38-54.

Chapter 7 The Anatomy of a Simulation Debriefing

Debriefing Modalities in Simulation and Operational Environments

Debriefing is a structured process that facilitates learning, reflection, and performance improvement following an event, training exercise, or real-world operation. Different debriefing modalities have been developed to accommodate various environments, learning preferences, and technological advancements. The four primary debriefing modalities are **self-reflection, self-debriefing, tele-debriefing, and facilitated debriefing**. Each modality has its distinct advantages and limitations, impacting how individuals and teams learn from their experiences.

Self-Reflection

Self-reflection is an internal, informal process in which individuals analyze their performance, thought processes, and decision-making after an event. This modality relies heavily on metacognition—the ability to think critically about one's thinking and actions (Schön, 1983). Professionals often use self-reflection in high-pressure environments, such as military operations or medical crises to assess their decision-making and emotional responses post-event (Kolb, 2014).

Studies have demonstrated that self-reflection can enhance problem-solving skills and situational awareness by encouraging individuals to recognize cognitive biases and adjust future actions accordingly (Dewey, 1933). However, one limitation of self-reflection is the absence of external validation, which can reinforce incorrect assumptions or an incomplete understanding of performance gaps (Fanning & Gaba, 2007).

For instance, a trauma surgeon who self-reflects after a complex procedure may realize they hesitated in deciding but may not recognize whether the hesitation was warranted without peer feedback or objective performance metrics.

Advantages:

- Encourages critical thinking and self-awareness
- Can be done immediately without external resources
- Promotes personal accountability and continuous learning

Limitations:

- Susceptible to cognitive biases and incomplete self-assessment
- Lacks external feedback to validate conclusions

Self-Debriefing

Self-debriefing is a structured, independent review process in which individuals evaluate their performance using predetermined frameworks, written reflections, or debriefing checklists. Unlike self-reflection, self-debriefing is **more structured and intentional**, often guided by specific questions or rubrics (Sawyer et al., 2016).

Research indicates that self-debriefing is most effective when combined with facilitated debriefing at a later stage, as it allows participants to organize their thoughts before engaging in team discussions (Rudolph et al., 2006). In aviation, pilots frequently complete self-debriefing reports before engaging in after-action reviews with their teams (Salas et al., 2008).

A study on nursing education found that self-debriefing using structured reflection checklists improved clinical decision-making and increased retention of key skills compared to unstructured debriefing methods (Cheng et al., 2016). However, one challenge of self-debriefing is that individuals may lack the expertise to recognize critical errors without external input, limiting its effectiveness in high-risk professions.

Advantages:

- Provides a structured reflection process
- Enhances retention of key learning points
- Can be integrated with facilitated debriefing for deeper insights

Limitations:

- Less effective without expert feedback or peer discussion
- Individuals may overlook critical errors or misinterpret events

Tele-Debriefing

Tele-debriefing involves remote debriefing sessions using video conferencing, online collaboration tools, or asynchronous discussion platforms. It has gained popularity due to the expansion of virtual learning environments and the need for geographically dispersed teams to conduct structured debriefings (McCoy et al., 2017).

Tele-debriefing is widely used in military operations, telemedicine, and aviation training, where experts may not be physically present to facilitate a debriefing session. For instance, remote surgical teams participating in robotic-assisted procedures use tele-debriefing to discuss case outcomes and improve technique standardization (Ahmed et al., 2013).

Studies have shown that tele-debriefing can be as effective as in-person debriefing if well-structured and supported by strong facilitation (Dieckmann et al., 2009). However, challenges such as reduced non-verbal communication, technology barriers, and participant engagement must be carefully managed (Kolbe et al., 2013).

Advantages:

- Increases accessibility for remote teams
- Allows for expert participation without geographic limitations
- Facilitates consistent training across multiple locations

Limitations:

- Loss of non-verbal cues and real-time interaction nuances
- Technical issues can disrupt engagement and learning
- Requires strong facilitation to maintain discussion quality

Facilitated Debriefing

Facilitated debriefing is the **gold standard** for performance improvement. It offers a structured, guided discussion led by an experienced facilitator. This method is used across **healthcare, military, aviation, and emergency response** to ensure that lessons learned from simulations, missions, or real-life events translate into actionable improvements (Raemer et al., 2011).

Facilitated debriefing often follows evidence-based models such as:

- **Plus/Delta**
- **PEARLS** (Promoting Excellence and Reflective Learning in Simulation) (Eppich & Cheng, 2015)
- **After-action reviews (AARs)** used by the U.S. military (U.S. Army, 1984)
- **Advocacy-Inquiry Model**, which balances observation with participant reflection (Rudolph et al., 2006)

Research suggests that facilitated debriefing significantly improves team performance, reduces medical errors, and enhances communication skills (Fanning & Gaba, 2007). In healthcare simulation, teams that participated in facilitated debriefings demonstrated a 30% improvement in crisis resource management skills compared to those who engaged in self-directed reviews alone (Cheng et al., 2014).

Advantages:

- Provides expert guidance and structured learning
- Enhances team communication and collaboration
- Identifies systemic issues and areas for improvement

Limitations:

- Requires trained facilitators, which may not always be available
- More time-intensive compared to self-directed modalities

Each debriefing modality offers unique benefits and challenges, making it critical to select the appropriate approach based on the environment, goals, and resources available. While self-reflection and self-debriefing foster individual learning, tele-debriefing and facilitated debriefing enhance team communication and collective performance. Research consistently supports the value of facilitated debriefing as the most effective modality, yet combining multiple approaches—such as self-debriefing followed by expert-led discussions—can further optimize learning outcomes.

By integrating these modalities into training programs and operational workflows, organizations can enhance decision-making, resilience, and overall performance, ultimately improving outcomes in both military and civilian domains.

Phases: Reaction, Analysis, and Application

Simulation debriefing is a structured process that enables participants to reflect on their experiences, analyze their actions, and apply lessons learned to future scenarios. Effective debriefings follow a phased approach, typically divided into three stages: reaction, analysis, and application. Each phase serves a specific purpose and builds upon the previous one to maximize learning outcomes.

Reaction Phase

The reaction phase focuses on participants' immediate emotional responses to the simulation. This stage allows individuals to express their thoughts and feelings, setting the tone for the rest of the debriefing. Facilitators often begin by asking open-ended questions like, "How do you feel the simulation went?" or "What stood out to you?" This phase is crucial for helping participants process any stress or anxiety they may have experienced during the simulation.

For example, in a high-fidelity simulation of a cardiac arrest, a nurse might express feeling overwhelmed by the rapid sequence of events. Another participant might note feeling unsure of their role within the team. These reflections provide insight into the emotional impact of the scenario but also highlight areas for further discussion. By acknowledging and normalizing these emotions, facilitators create an atmosphere of psychological safety, encouraging participants to engage openly in the subsequent phases.

Analysis Phase

The analysis phase is the core of the debriefing process. In this stage, participants delve into the specifics of what happened during the simulation, exploring both successes and areas for improvement. Facilitators guide the discussion with targeted questions, prompting participants to reflect on their decision-making, communication, and teamwork.

A common technique in this phase is the "Advocacy-Inquiry" method, where facilitators balance observations with questions. For instance, a facilitator might say, "I noticed a delay in initiating chest compressions. Can you walk me through what was happening at that moment?" This approach encourages participants to think critically about their actions while avoiding blame.

The analysis phase often incorporates video playback or objective data, such as time-to-intervention metrics, to enhance understanding. In one debriefing, a team reviewing a trauma scenario might use video footage to analyze how well they followed the ABC (airway, breathing, circulation) protocol. Discussions during this phase often reveal hidden dynamics, such as unspoken assumptions or miscommunications, that can profoundly affect team performance.

For example, in a neonatal resuscitation simulation, the debrief might uncover that while the team adhered to clinical guidelines, a lack of role clarity led to redundant efforts and delays in critical interventions. By identifying these gaps, participants can collaboratively develop strategies to optimize performance in future scenarios.

Application Phase

The application phase focuses on translating insights from the debriefing into actionable improvements. Participants identify key takeaways and develop strategies to apply these lessons in real-world settings. Facilitators might ask questions like, "What will you do differently next time?" or "How can we integrate these lessons into our daily practice?"

For example, after a simulation of a difficult airway management case, a team might implement a checklist to improve role clarity during emergencies. Another team, reflecting on a delayed diagnosis in a sepsis scenario, might commit to adopting an early warning scoring system. This phase ensures that the learning from the simulation extends beyond the training environment, fostering continuous improvement in clinical practice.

Importance of Structured Frameworks

Structured frameworks provide a foundation for effective debriefing, ensuring that discussions remain focused, constructive, and goal oriented. Four widely used frameworks in simulation debriefing are the "Plus-Delta," "Advocacy-Inquiry," GAS, and PEARLS models.

Plus-Delta Framework

The Plus-Delta framework is a straightforward tool that encourages participants to identify what went well (pluses) and what could be improved (deltas). This simple structure ensures that debriefings are balanced and actionable. For instance, after a simulation of a sepsis case, a team might identify clear communication during handoffs as a plus, noting that antibiotic administration delays were a delta. This framework's simplicity makes it especially useful for teams new to debriefing, as it provides a clear and accessible starting point for discussion.

Advocacy-Inquiry Model

The Advocacy-Inquiry model combines observation with open-ended questioning to foster critical reflection. Facilitators begin by sharing an observation (advocacy) and then ask a question (inquiry) to explore the reasoning behind participants' actions. This method promotes deeper analysis and avoids defensiveness. For example, a facilitator might say, "I noticed there was hesitation in administering epinephrine. Can you share what factors influenced that decision?"

The advocacy-inquiry model fosters a culture of curiosity and continuous learning by encouraging participants to explore the reasoning behind their actions. Over time, this approach helps teams build trust and develop a shared commitment to improvement.

GAS Model

The Gather-Analyze-Summarize (GAS) model is a structured debriefing framework designed to guide participants through a reflective process. The "Gather" phase focuses on collecting initial reactions and observations from participants, similar to the reaction phase of debriefing. The "Analyze" phase involves a detailed discussion of what occurred during the simulation, emphasizing critical incidents and their underlying causes. Finally, the "Summarize" phase consolidates key insights and actionable steps to improve future performance. The GAS model's linear structure provides a clear roadmap for facilitators, ensuring thorough and cohesive debriefings.

PEARLS Model

The Promoting Excellence and Reflective Learning in Simulation (PEARLS) model is a structured yet flexible debriefing framework designed to maximize learning by blending learner-centered and instructor-led techniques. Developed by Eppich and Cheng (2015), this model recognizes that debriefing is not a one-size-fits-all process and that different scenarios, participants, and learning objectives require different facilitation strategies. Unlike rigid debriefing models that follow a single approach, PEARLS encourages facilitators to adapt their methods in real-time, ensuring that discussions remain engaging, educational, and aligned with learners' needs.

Core Principles of the PEARLS Model

PEARLS is built on the understanding that debriefing must be dynamic to accommodate participants' varying levels of experience, knowledge, and emotional responses. The model integrates three key debriefing approaches, which facilitators can switch between depending on the situation:

- Direct Instruction: The facilitator provides specific guidance or corrects misunderstandings when learners require explicit feedback on clinical knowledge, procedural skills, or technical aspects of performance.

- Facilitated Discussion: Participants engage in an open-ended conversation to reflect on their performance, guided by the facilitator's questions. This approach encourages self-discovery and critical thinking.

- Self-Reflection: The facilitator prompts individuals or teams to analyze their decision-making processes, helping them uncover assumptions and cognitive biases.

The PEARLS model creates an optimal balance between structure and learner autonomy by incorporating elements of all three approaches.

Applying PEARLS in Simulation-Based Learning

One of the key strengths of PEARLS is its adaptability to different learning contexts. Consider a healthcare simulation where a team of emergency medicine residents responds to a simulated code blue scenario involving a cardiac arrest patient. If the team miscalculates the medication dosage, the facilitator must decide how best to address the error during the debriefing. Using the PEARLS model, they might employ direct instruction to correct the mistake by reviewing the proper dosing guidelines. However, rather than relying solely on instruction, the facilitator could transition to a facilitated discussion, prompting participants to reflect on what led to the miscalculation—perhaps a breakdown in communication or cognitive overload during the high-stress event. The session could then conclude with self-reflection, encouraging team members to consider better managing similar situations.

The Flexibility of PEARLS in Different Contexts

The versatility of PEARLS makes it an effective debriefing tool across diverse settings, including healthcare, military training, aviation, and corporate crisis management. For example, in a military after-action review following a field training exercise, the instructor might start with direct instruction, explaining key tactical errors, then move to a group discussion to explore decision-making under pressure. Similarly, in an aviation training scenario, pilots might self-reflect on how they handled an emergency landing before discussing alternative strategies in a facilitated discussion.

This adaptability ensures that learning objectives remain the central focus while accommodating different learning styles and team dynamics. Research has shown that when learners are actively engaged in the debriefing process—rather than passively receiving information—they are more likely to retain knowledge, develop critical thinking skills, and improve team communication (Eppich & Cheng, 2015).

Why PEARLS is an Effective Debriefing Model

Several factors make the PEARLS model particularly effective:

- Customizable to Learner Needs: The facilitator can tailor debriefing sessions based on the learners' experience levels, cognitive load, and emotional responses to the scenario.
- Balances Structure and Flexibility: The model provides a clear debriefing framework, allowing facilitators to pivot as needed.
- Enhances Psychological Safety: By incorporating learner-centered discussions and self-reflection, PEARLS fosters an environment where participants feel comfortable analyzing their mistakes without fear of judgment.
- Promotes Deeper Learning: The combination of direct feedback and guided discovery ensures learners internalize lessons more effectively than lecture-based debriefing alone.

The PEARLS model represents an evolution in debriefing methodology, shifting from a rigid one-way feedback process to a dynamic, learner-driven experience. This approach accommodates diverse learning needs by integrating direct instruction, facilitated discussion, and self-reflection while maintaining a structured debriefing framework. Whether applied in healthcare, military, aviation, or corporate crisis training, PEARLS fosters critical thinking, teamwork, and continuous performance improvement. Its blended approach ensures that debriefing remains a powerful tool for individual skill development and collective team learning, ultimately leading to better decision-making and improved outcomes in high-stakes environments.

The Role of Adult Learning and Uncovering Frames

Adult learning principles are central to effective debriefing because simulation participants are often experienced professionals or advanced learners. Adults learn best when they can actively engage with the material, reflect on their experiences, and see the relevance of new knowledge to their daily practice. Debriefing aligns with these principles by providing a structured opportunity for self-directed exploration and meaningful feedback.

One key debriefing component is uncovering frames—the mental models, assumptions, and thought processes influencing participants' actions. By exploring these frames, facilitators help learners understand the reasoning behind their decisions and identify opportunities for improvement. For instance, a team may delay administering medication during a simulated emergency because they assume the patient's condition is stable. Uncovering this frame allows the team to recognize the need for more accurate assessment techniques in real-world scenarios.

The Advocacy-Inquiry model is particularly effective in uncovering frames. For example, a facilitator might say, "I noticed hesitation in initiating advanced airway management. Can you share what you were thinking at that moment?" This approach reveals underlying assumptions and encourages participants to examine their cognitive processes without fear of judgment.

Facilitators must create a safe and supportive environment for uncovering frames, as this process often involves confronting deeply held beliefs or recognizing gaps in knowledge. By validating participants' experiences and reframing challenges as opportunities for growth, facilitators can foster a culture of curiosity and continuous learning.

Psychological Safety as a Foundation for Effective Debriefing

Psychological safety is the cornerstone of successful debriefing. Without it, participants may hesitate to engage fully, fearing judgment or blame. Creating a psychologically safe environment requires facilitators to set a tone of respect, openness, and support from the outset.

One key strategy is establishing ground rules, such as "What is said here stays here" and "Focus on behaviors, not individuals." Facilitators also play a critical role in modeling vulnerability and empathy. For example, a facilitator might share their experience with a similar challenge, demonstrating that mistakes are learning opportunities.

Psychological safety not only encourages openness but also enhances emotional resilience. In a debriefing after a high-stakes neonatal resuscitation simulation, participants might express feelings of frustration or guilt about perceived mistakes. By validating these emotions and reframing them as part of the learning process, facilitators help participants build confidence and adaptability.

Relevant Statistics

Compelling statistics support the impact of debriefing in medical and nursing education:

- Research indicates that structured debriefing can improve clinical performance by up to **25%** in simulated settings (Cheng et al., 2014).
- A systematic review found that debriefing increases team communication and collaboration by **30%**, with corresponding improvements in patient safety outcomes (Kolbe et al., 2013).
- Studies in nursing education show that students participating in simulation debriefings demonstrate a **20%** higher retention of key clinical skills than those receiving traditional instruction alone (Liaw et al., 2016).
- Healthcare teams that engage in regular debriefing after simulations are **40%** more likely to identify latent safety threats, such as workflow inefficiencies and equipment issues, than teams without debriefing protocols (Fanning & Gaba, 2007).

Debriefings Are Not…

- A blame game
- A substitute for incident reporting
- A formal investigation
- A venting session
- Solely about negative events
- Just for leadership
- Optional in high-stakes settings

- Limited to immediate reaction

Key Takeaways

- **Four Primary Debriefing Modalities Enhance Learning and Performance**
    - **Self-reflection** promotes personal insight but lacks external validation.
    - **Self-debriefing** provides a structured review but is most effective when supplemented with peer feedback.
    - **Tele-debriefing** enables remote teams to analyze performance, though it requires strong facilitation.
    - **Facilitated debriefing** is the gold standard, offering expert-led guidance for in-depth analysis and team learning.

- **The Three-Phase Debriefing Process Maximizes Retention and Improvement**
    - The **Reaction Phase** allows participants to express emotions and establish psychological safety.
    - The **Analysis Phase** facilitates critical reflection on decisions, actions, and team dynamics.
    - The **Application Phase** translates insights into actionable improvements for real-world practice.

- **Structured Debriefing Frameworks Ensure Effective Reflection and Learning**

    - **Plus-Delta** highlights successes and areas for improvement.
    - **Advocacy-Inquiry** balances facilitator observations with open-ended questions to deepen learning.
    - **The GAS Model** (Gather, Analyze, Summarize) provides a structured discussion flow.
    - **PEARLS** (Promoting Excellence and Reflective Learning in Simulation) blends direct instruction with learner-centered reflection.

- **Psychological Safety is Essential for Meaningful Debriefing**

  - Facilitators must create a non-judgmental environment to encourage honest discussion.
  - Establishing ground rules, such as "focus on behaviors, not individuals," prevents blame and promotes constructive feedback.
  - Psychological safety increases emotional resilience, helping participants embrace learning from mistakes.

- **Debriefing Significantly Improves Clinical Skills, Communication, and Patient Safety**
  - Studies show **structured debriefing improves clinical performance by up to 25%** (Cheng et al., 2014).
  - Teams engaging in debriefing demonstrate **30% better communication and collaboration** (Kolbe et al., 2013).
  - **Simulation debriefing enhances retention of key clinical skills by 20%**, reinforcing its role in medical and nursing education (Liaw et al., 2016).
  - Teams that conduct regular debriefings are **40% more likely to identify safety threats**, reducing errors and improving patient outcomes (Fanning & Gaba, 2007).

Reflection

The anatomy of a simulation debriefing reveals the depth and complexity of this practice, highlighting its potential to transform learning and performance. By following a structured approach and prioritizing psychological safety, facilitators can create an environment where participants feel empowered to reflect critically and grow professionally. Reflecting on these principles encourages us to consider how they might be applied in our contexts, fostering a culture of continuous improvement and collaboration.

As simulation debriefing continues to evolve, its principles remind us of the power of structured reflection and feedback to drive innovation and resilience in any field. The stories of teams discovering hidden strengths, uncovering blind spots, and committing to change exemplify this practice's transformative potential.

References

Ahmed, K., Jawad, M., Abboudi, M., et al. (2013). Effectiveness of surgical tele-debriefing on technical performance. *The American Journal of Surgery, 205*(1), 50-55.

Cheng, A., Grant, V., Dieckmann, P., Arora, S., Robinson, T., & Eppich, W. (2014). Debriefing for technology-enhanced simulation: A systematic review and meta-analysis. *Medical Education, 48*(7), 657-666.

Cheng, A., Grant, V., Dieckmann, P., Arora, S., Robinson, T., & Eppich, W. (2016). Debriefing for technology-enhanced simulation: A systematic review and meta-analysis. *Medical Education, 50*(4), 469-478.

Dieckmann, P., Gaba, D., & Rall, M. (2009). Deepening the theoretical foundations of patient simulation as social practice. *Simulation in Healthcare, 4*(3), 183-193.

Eppich, W., & Cheng, A. (2015). Promoting excellence and reflective learning in simulation (PEARLS). *Simulation in Healthcare, 10*(2), 106-115.

Eppich, W., & Cheng, A. (2015). Promoting excellence and reflective learning in simulation (PEARLS): Development and rationale for a blended approach to health care simulation debriefing. *Simulation in Healthcare, 10*(2), 106-115.

Fanning, R. M., & Gaba, D. M. (2007). The role of debriefing in simulation-based learning. *Simulation in Healthcare, 2*(2), 115-125.

Kolb, D. A. (2014). *Experiential learning: Experience as the source of learning and development* (2nd ed.). Pearson Education.

Kolbe, M., Grande, B., & Spahn, D. R. (2013). The ABC of debriefing: A systematic approach to feedback in simulation. *Advances in Simulation, 2*(1), 2-12.

Liaw, S. Y., Zhou, W. T., Lau, T. C., Siau, C., & Chan, S. W. (2016). An interprofessional communication training using simulation to enhance safe care for a deteriorating patient. *Nurse Education Today, 38*, 21-27.

Schön, D. A. (1983). *The Reflective Practitioner: How Professionals Think in Action.* Basic Books.

U.S. Army. (1984). *After-Action Reviews: Guidelines and Techniques.* U.S. Government Printing Office.

Chapter 8
Adapting to the Healthcare Context

Differences Between Military and Healthcare Environments

While military debriefing practices have significantly influenced healthcare, notable differences between the two environments require adaptations to ensure successful implementation. These differences center on hierarchy, organizational culture, and operational dynamics.

Hierarchy and Power Dynamics

In military settings, rank and structure are deeply ingrained, and the chain of command often dictates decision-making and communication. Debriefings in this context tend to emphasize collective accountability, with leaders modeling vulnerability to encourage open dialogue.

Healthcare teams, while also hierarchical, often face additional challenges due to the diverse professional backgrounds of team members. Physicians, nurses, technicians, and other healthcare professionals each bring unique expertise and perspectives, but power dynamics can inhibit open communication. For instance, a nurse might hesitate to voice concerns about a physician's decision during a debriefing, fearing repercussions or judgment.

To address this, healthcare debriefings must prioritize psychological safety and inclusivity. Facilitators play a critical role in leveling the playing field, ensuring that all voices are heard and valued. Strategies such as establishing ground rules and using structured frameworks like Advocacy-Inquiry can help mitigate the effects of hierarchy. One example is a hospital that implemented pre-briefing sessions to set expectations, ensuring all participants felt prepared and empowered to engage in the debriefing process.

Organizational Culture

The military's culture is characterized by discipline, structure, and a shared commitment to mission objectives. Debriefing practices in this environment align naturally with these cultural norms, emphasizing continuous improvement and resilience.

Healthcare culture, by contrast, is more heterogeneous. Institutions vary widely in their approach to teamwork, communication, and accountability. Some organizations may have a strong emphasis on collaborative learning, while others operate in silos. This variability requires debriefing practices to be tailored to the specific cultural context of each institution.

For example, in a hospital with a historically top-down approach to decision-making, facilitators may need to invest extra effort in fostering psychological safety and encouraging open dialogue. Conversely, in organizations with a strong culture of teamwork, debriefings may naturally align with existing practices, requiring less adaptation. One case study highlights how a healthcare institution shifted its culture over time by incorporating regular debriefings into daily workflows, creating a ripple effect that improved overall team cohesion.

Operational Dynamics

Military operations are often high-stakes, time-sensitive, and conducted in highly controlled environments. This predictability enables the integration of standardized debriefing protocols.

While similarly high-stakes, healthcare is less predictable due to the complex and dynamic nature of patient care. Emergencies can arise without warning, requiring immediate responses that leave little time for formal debriefing. To overcome this challenge, healthcare teams can incorporate brief, informal "hot washes" immediately after critical events, followed by more detailed debriefings. This two-tiered approach ensures that key insights are captured without disrupting patient care.

For instance, an emergency department might conduct a quick five-minute discussion after resuscitation to identify immediate takeaways, such as equipment readiness or communication barriers. Later, a more comprehensive debriefing involving the entire team could delve deeper into system-level improvements.

Aspect	Military Environment	Healthcare Environment
Hierarchy	Strict, rank-based structure with clear chains of command.	Diverse professional hierarchies; power dynamics can vary between physicians, nurses, technicians, and others.
Communication	Focused on standard operating procedures and explicit communication.	Requires bridging interprofessional communication gaps; often lacks standardized protocols across all roles.
Culture	Emphasizes discipline, structure, and collective mission goals.	Varies significantly; some institutions prioritize collaboration, while others operate in silos.
Operational Dynamics	High-stakes, time-sensitive, and controlled environments with predictable workflows.	High-stakes but less predictable; patient care is dynamic, with emergencies arising unexpectedly.
Debriefing Integration	Standardized protocols and routine incorporation into operations.	Requires flexible approaches, such as "hot washes" for immediate feedback and formal sessions for in-depth analysis.
Team Composition	Homogeneous teams with uniform training and shared mission objectives.	Heterogeneous teams with diverse expertise, training, and perspectives.

Aspect	Military Environment	Healthcare Environment
Psychological Safety	Leaders often model vulnerability to encourage open dialogue.	Psychological safety must be actively cultivated to address power imbalances and encourage inclusivity.

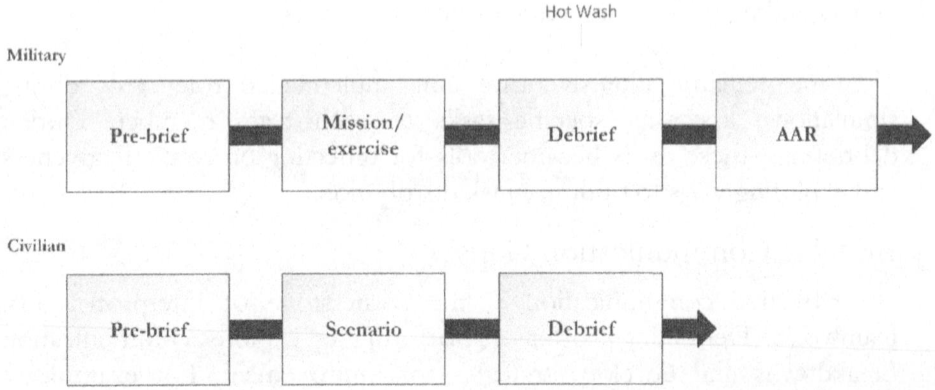

Tailoring Debriefing to Interprofessional Teams

Healthcare teams are inherently interprofessional, comprising individuals from diverse disciplines with varying levels of expertise and training. This diversity is a strength, but it also presents challenges for debriefing.

Emphasizing Inclusivity

Interprofessional debriefings must ensure that all participants feel empowered to contribute. Facilitators should encourage active participation from everyone, using techniques like round-robin discussions or direct invitations to share perspectives. For example, a facilitator might say, "Let's hear from the respiratory therapist about their experience during the intubation process."

Inclusivity enriches the discussion and fosters mutual respect among team members. In one scenario, a surgical team discovered that input from a scrub nurse led to the identification of a critical workflow inefficiency, highlighting the value of diverse perspectives.

Addressing Role Clarity

Role clarity is essential in interprofessional teams, particularly during simulations or critical events. Debriefings provide an opportunity to discuss how roles were defined and executed, identifying gaps or overlaps that may have impacted performance. For instance, during a debriefing, a team might discover that two members assumed responsibility for medication administration, leading to delays. Addressing these issues fosters a clearer understanding of roles and responsibilities.

One pediatric intensive care unit implemented role cards during simulations, assigning specific tasks to each team member. During debriefings, these cards became tools for reflecting on role effectiveness and exploring ways to optimize task distribution.

Bridging Communication Gaps

Effective communication is the cornerstone of interprofessional teamwork. Debriefings offer a platform to explore communication breakdowns and develop strategies for improvement. For example, a debriefing might reveal that a lack of closed-loop communication during a code blue scenario led to missed steps. By practicing techniques like repeating back orders, teams can enhance communication and reduce errors.

One notable case involved a hospital team that struggled with handoffs between shifts. Through regular debriefings, they identified inconsistencies in the information shared during transitions, leading to the development of a standardized handoff checklist. This change improved communication and reduced errors related to incomplete information transfer.

Academic and Clinical Debriefing

Debriefing is an essential practice in both academic and clinical settings, designed to enhance learning and improve outcomes. However, both types of debriefing share common goals—such as reflecting on performance and identifying areas for improvement—but their processes, applications, and contextual nuances differ significantly. Understanding these differences can help educators, healthcare professionals, and administrators tailor their approaches to meet specific learning objectives and practice standards.

Academic Debriefing

Academic debriefing occurs primarily in educational environments like simulation labs, classrooms, or training programs. It is structured to facilitate learning by encouraging participants to reflect on their actions, analyze outcomes, and connect theory to practice. Academic debriefing typically follows simulation exercises or controlled scenarios to replicate real-life situations.

Key Features of Academic Debriefing:

- **Structured Frameworks:** Academic debriefings often use established models, such as the Debriefing for Meaningful Learning (DML) framework or the "PEARLS" (Promoting Excellence and Reflective Learning in Simulation) method. These frameworks provide a systematic approach to ensure participants achieve educational objectives.
- **Focus on Learning Objectives:** The primary goal is to align participants' reflections with pre-defined learning objectives, such as developing clinical reasoning, teamwork, or technical skills.
- **Facilitator Role:** Facilitators in academic debriefing are typically instructors or simulation educators trained in guiding discussions and fostering a psychologically safe environment for learners to discuss errors and successes openly.
- **Feedback and Reflection:** Emphasis is placed on guided reflection and constructive feedback, allowing participants to identify areas for growth while receiving insights from facilitators.

Clinical Debriefing

Clinical debriefing occurs in real-world healthcare settings, often immediately following a clinical event, procedure, or critical incident. The purpose is to improve patient care, enhance team communication, and address any psychological stress experienced by healthcare providers. Unlike academic debriefing, which is primarily educational, clinical debriefing often has an operational or emotional focus.

Key Features of Clinical Debriefing:

- **Situational Relevance:** Clinical debriefings address real-time events and identify immediate lessons to improve future practice and patient safety.
- **Emotional Support:** Given the high-stakes and emotionally charged nature of clinical environments, clinical debriefing often incorporates an element of emotional processing to address stress or trauma experienced by team members.
- **Adaptive Structure:** While clinical debriefing can follow structured models like the "LEARN" framework (Listen, Empathize, Analyze, Reflect, Next Steps), it is often less rigid to accommodate the dynamic nature of clinical practice.
- **Multidisciplinary Participation:** These debriefings frequently involve diverse healthcare team members, fostering interprofessional collaboration and communication.

Key Comparisons

Aspect	Academic Debriefing	Clinical Debriefing
Setting	Educational environments, such as simulation labs	Real-world healthcare settings
Purpose	Facilitating learning and skill development	Improving patient care, team dynamics, and emotional well-being
Structure	Highly structured frameworks	Flexible, context-dependent
Focus	Pre-defined learning objectives	Immediate lessons and emotional processing
Facilitators	Educators or simulation trainers	Clinicians, team leaders, or peer facilitators
Participants	Learners (students, trainees)	Multidisciplinary healthcare teams

Integrating Academic and Clinical Debriefing

Although academic and clinical debriefing differ in focus and application, cross-pollination has significant potential. For example, academic debriefing techniques can be adapted to clinical settings to enhance structure and learning outcomes. Conversely, the emotional and situational awareness integral to clinical debriefing can inform more holistic approaches in academic environments. By combining the strengths of both methods, organizations can foster a culture of continuous learning and improvement.

Example of End of Shift Debrief/Post-conference for a Group

Elements	Example
1. Introduction 2. Reactions 3. Discussion 4. Closing	1. *We will spend 10-20 minutes completing this post-conference. If you still have questions/ concerns after this debrief, I will be around after we dismiss the group.* *Any patient information discussed is considered privileged and will follow HIPAA rules. All report sheets or notes with patient information needs to be put in the shred box as you leave.* *I would like to go around the table and discuss their reflection of your shift, what you think went well, and what you thought could have been done differently or an educational gap you've identified.* 2. [Invite sharing of emotions and reactions] Listen, affirm, and validate. 3. [Pluses] *What went well?* [Deltas] *What could be done differently or improve?* [Educational Gaps Identified] *What educational gaps did you come across today and how do you plan to close that gap?* 4. *To close the session, lets go around the table and tell me a "take-away" you got from this rotation.* *Thank you for your participation.*

Adapted from Center for Medical Simulation

The Role of Evidence-Based Debriefing Practices

Evidence-based debriefing practices provide a foundation for effective learning and performance improvement in healthcare. Research has shown that structured debriefings can enhance team communication, reduce errors, and improve patient outcomes.

Incorporating Research-Backed Frameworks

Frameworks such as Plus-Delta, Advocacy-Inquiry, GAS, and PEARLS are rooted in evidence and have been proven to facilitate effective debriefing. For example, a systematic review found that teams using structured debriefing frameworks identified 40% more latent safety threats than teams without such frameworks (Cheng et al., 2014). By adopting these models, healthcare teams can ensure consistent and impactful debriefings.

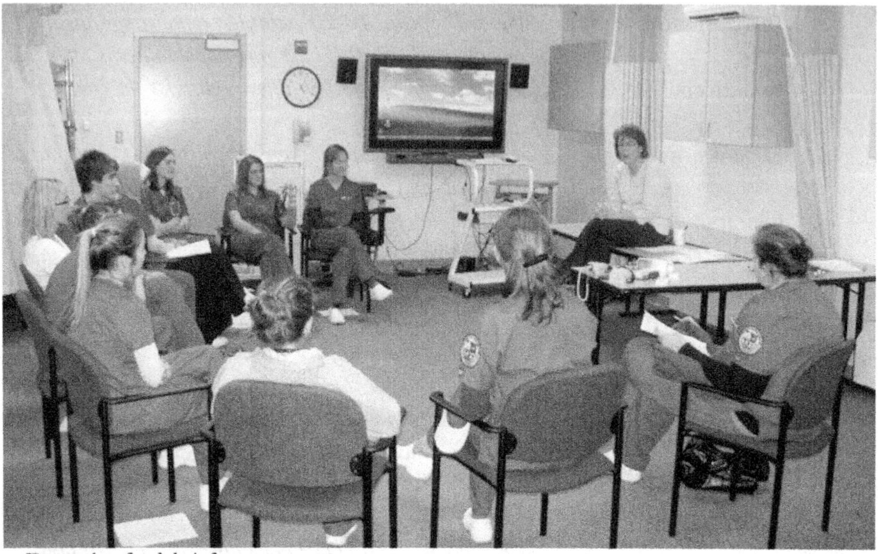

Example of a debrief.
Photograph courtesy of Healthsimualtion.com

Measuring Impact

Debriefing practices should be regularly evaluated to ensure their effectiveness. Metrics such as team performance, communication scores, and patient safety indicators can provide valuable insights. For instance, a hospital implementing structured debriefings in its emergency department might track the time-to-intervention for critical procedures before and after introducing debriefing practices. Improvements in these metrics can validate the effectiveness of the debriefing process and justify its continued use.

Adapting to Emerging Evidence

As healthcare continues to evolve, so too must debriefing practices. Emerging research may highlight new strategies or tools that can enhance debriefing effectiveness. For example, integrating technology, such as video-assisted debriefing, has shown promise in providing more detailed and objective feedback. Staying abreast of these advancements ensures that debriefing remains a cutting-edge tool for learning and improvement.

Key Takeaways

- Adapting military debriefing practices to healthcare requires addressing differences in hierarchy, culture, and operational dynamics.
- Interprofessional debriefings must emphasize inclusivity, role clarity, and communication to maximize their impact.
- Evidence-based frameworks, such as Plus-Delta, Advocacy-Inquiry, GAS, and PEARLS, provide a foundation for effective debriefing in healthcare.
- Regular evaluation and adaptation of debriefing practices ensure their continued relevance and effectiveness in improving patient care.

Reflection

Adapting debriefing practices to healthcare highlights the importance of flexibility, inclusivity, and a commitment to continuous improvement. By addressing the unique challenges of the healthcare environment, we can unlock the full potential of debriefing as a tool for learning and performance enhancement. Reflecting on these adaptations reminds us that the success of debriefing lies not only in its structure but also in its ability to bring diverse teams together to achieve shared goals.

References

Cheng, A., Grant, V., Dieckmann, P., Arora, S., Robinson, T., & Eppich, W. (2014). Debriefing for technology-enhanced simulation: A systematic review and meta-analysis. *Medical Education*, *48*(7), 657-666.

Eppich, W., & Cheng, A. (2015). Promoting excellence and reflective learning in simulation (PEARLS): Development and rationale for a blended approach to health care simulation debriefing. *Simulation in Healthcare*, *10*(2), 106-115.

Fanning, R. M., & Gaba, D. M. (2007). The role of debriefing in simulation-based learning. *Simulation in Healthcare*, *2*(2), 115-125.

Kolbe, M., Grande, B., & Spahn, D. R. (2013). The ABC of debriefing: A systematic approach to feedback in simulation. *Advances in Simulation*, *2*(1), 2-12.

Liaw, S. Y., Zhou, W. T., Lau, T. C., Siau, C., & Chan, S. W. (2016). An interprofessional communication training using simulation to enhance safe care for a deteriorating patient. *Nurse Education Today*, *38*, 21-27.

Chapter 9
Timing of Debriefing and Time Allocation

Introduction

Debriefing is a cornerstone of simulation-based learning, allowing one to reflect, analyze, and derive meaning from the experience. Its timing and duration significantly impact its effectiveness, influencing how well participants internalize the lessons learned. Picture a group of healthcare providers who have just completed a high-pressure simulation of a cardiac arrest scenario. The adrenaline is still pumping, and questions swirl in their minds: "Did I respond quickly enough?" "Could I have communicated more effectively with the team?" This is when debriefing becomes critical. It's a chance to unpack actions, decisions, and emotions while still vivid.

This chapter explores the optimal timing for debriefing, recommended time allocation, and the factors influencing these decisions, supported by research and industry guidelines. By understanding the nuances of debriefing timing and structure, facilitators can enhance learning outcomes and foster a culture of continuous improvement.

Timing of Debriefing

Immediate vs. Delayed Debriefing

The debate between immediate and delayed debriefing often centers on the simulation's goals and the participants' readiness. Imagine finishing a challenging simulation only to be told, "We'll debrief tomorrow." For many participants, this delay risks losing the emotional and cognitive engagement vital to reflective learning. Research overwhelmingly supports the efficacy of immediate debriefing. Studies, such as those by Dieckmann et al. (2012), indicate that debriefing immediately after the simulation maximizes retention and accelerates applying new skills.

Advantages of Immediate Debriefing:

- Participants' emotional responses are fresh, allowing facilitators to address concerns and reinforce learning in real time.
- Details of the scenario remain vivid, making it easier to analyze actions and decisions.
- Immediate feedback enables participants to apply insights in subsequent simulations or real-world situations.

However, there are circumstances where delayed debriefing may be appropriate. For example, after a particularly intense or emotionally charged simulation, participants may benefit from a short period of reflection before engaging in a structured discussion. Similarly, logistical challenges, such as resetting equipment or accommodating varying schedules, may necessitate a delay.

Recommended Practice

Professional organizations, such as the Society for Simulation in Healthcare (SSH) and the International Nursing Association for Clinical Simulation and Learning (INACSL), advocate for immediate debriefing when possible. Immediate debriefing allows facilitators to build on participants' heightened engagement and emotional responses, ensuring that learning is meaningful and enduring.

Time Allocation for Debriefing

Industry Standards

Time allocation for debriefing is as much an art as a science. Imagine condoning a 30-minute simulation into a 5-minute debriefing—the result would likely feel rushed and superficial. On the other hand, an excessively long debriefing risks losing participants' focus and engagement. According to the INACSL and other industry leaders, effective debriefing typically lasts **20-40 minutes**, depending on the complexity of the scenario and the number of participants.

A general rule of thumb is to allocate debriefing time proportionally to the simulation duration. For shorter scenarios (under 20 minutes), the debriefing should be at least twice the length of the simulation. For longer simulations, the debriefing duration may match or slightly exceed the simulation time.

Factors Influencing Time Allocation

1. **Scenario Complexity:** Picture a simulation focused on a routine medication administration versus one involving a multi-trauma patient requiring rapid decision-making and team coordination. The former may require minimal debriefing, while the latter demands extensive discussion to unpack the intricate dynamics.
2. **Participant Group Size:** In smaller groups, participants have more opportunities to share insights, potentially shortening the debriefing time. Conversely, larger groups require additional time to ensure everyone has a voice.
3. **Facilitator Skill Level:** An experienced facilitator can guide discussions efficiently, ensuring that key learning objectives are addressed without veering off track. Conversely, less experienced facilitators may need more time to manage discussions effectively.
4. **Learning Objectives:** Technical skills, such as inserting an IV, may require a straightforward debrief, while scenarios focused on communication, teamwork, or ethical dilemmas often necessitate deeper exploration.
5. **Participant Experience Level:** Novices may need additional time to process feedback and understand complex concepts, while seasoned professionals can often engage in quicker, more focused discussions.

Suggested Debriefing Time Breakdown

To ensure a balanced and effective debriefing, facilitators can follow a structured framework:

Phase	Time Allocation	Description
Reaction Phase	5-10 minutes	Participants share immediate reactions and emotions.
Analysis Phase	10-20 minutes	The facilitator guides the discussion of decisions, actions, and outcomes.
Summary Phase	5-10 minutes	Key takeaways and action items are identified.

Statistical Insights

Data underscore the importance of timing and structure in debriefing. A survey conducted by the Healthcare Simulation Standards Consortium (2020) revealed that:

- **78%** of facilitators allocate 20-40 minutes for debriefing sessions.
- **85%** of participants reported higher satisfaction and learning outcomes when debriefing lasted at least 30 minutes.
- **72%** of facilitators noted that immediate debriefing improved participant engagement and learning retention.

These statistics highlight the need for thoughtful time allocation and reinforce the benefits of immediate, structured debriefing.

Challenges in Time Management

Despite the best intentions, time management during debriefing can be challenging. Facilitators may face:

- **Time Constraints:** Tight schedules or competing priorities can pressure facilitators to rush through debriefings, potentially undermining their effectiveness.
- **Overextended Discussions:** Without clear boundaries, discussions can veer off-topic, consuming valuable time without adding to the learning experience.

- **Participant Readiness:** Balancing the needs of participants who process quickly with those who require more time can be difficult.

Strategies for Effective Time Management

1. **Set Clear Objectives:** Begin by outlining the specific goals of the debriefing. This helps participants stay focused and ensures that key learning points are addressed.
2. **Use Structured Frameworks:** Models like the "Gather, Analyze, Summarize" (GAS) or PEARLS (Promoting Excellence and Reflective Learning in Simulation) provide clear guidance for debriefing.
3. **Monitor Time Closely:** Assign a timekeeper or use visual cues to ensure discussions stay on track and within the allocated timeframe.
4. **Engage Participants:** Use targeted questions to encourage meaningful contributions without making discussions redundant or tangential.
5. **Adapt as Needed:** Be prepared to adjust the debriefing duration based on the group's engagement and the scenario's complexity.

Key Takeaways

1. Immediate debriefing is generally more effective than delayed sessions, leveraging participants' emotional engagement and memory.
2. Industry standards recommend allocating 20-40 minutes for debriefing, adjusted for scenario complexity and group size.
3. A structured approach, clear objectives, and skilled facilitation are essential for maximizing the impact of debriefing.

Reflection

Consider how the timing and structure of your debriefing sessions influence participant engagement and learning outcomes. Are there opportunities to refine your approach, ensuring that debriefings are both efficient and impactful?

References

Dieckmann, P., Gaba, D., & Rall, M. (2012). Deepening the theoretical foundations of patient simulation as social practice. *Simulation in Healthcare, 7*(3), 183-193.

Healthcare Simulation Standards Consortium. (2020). Survey on debriefing practices and time allocation.

INACSL Standards Committee. (2016). INACSL standards of best practice: Simulation debriefing. *Clinical Simulation in Nursing, 12*(S), S21-S25.

Rudolph, J. W., Simon, R., Dufresne, R. L., & Raemer, D. B. (2006). There's no such thing as "nonjudgmental" debriefing: A theory and method for debriefing with good judgment. *Simulation in Healthcare, 1*(1), 49-55.

Chapter 10
Traditional Debriefing vs. Video Debriefing in Healthcare Simulation

Introduction

Debriefing is integral to simulation-based education, allowing participants to reflect on their actions, analyze outcomes, and integrate lessons into practice. Traditionally, debriefing has relied on facilitated discussions, but technological advancements have introduced video debriefing as a powerful alternative. Imagine a team of healthcare professionals huddled in a simulation room after managing a simulated cardiac arrest. The facilitator asks, "What went well, and what could we improve?" This traditional approach emphasizes real-time reflection. Now, picture the same team reviewing video footage of their simulation, analyzing their communication, teamwork, and decision-making. These two methods, while distinct, offer unique benefits and challenges.

This chapter explores traditional and video debriefing, comparing their methodologies, benefits, limitations, and value in simulation-based learning. By understanding the strengths and nuances of each approach, educators can tailor their debriefing strategies to maximize learning outcomes.

Traditional Debriefing

Traditional debriefing is a discussion-based process by a facilitator immediately after a simulation. It focuses on participants' recollections, emotions, and insights and is guided by structured frameworks like the "Gather, Analyze, Summarize" (GAS) model or the PEARLS (Promoting Excellence and Reflective Learning in Simulation) approach.

Key Characteristics

1. **Facilitator-Led Discussion:** A skilled facilitator takes center stage, guiding participants through a reflective process that uncovers both successes and areas for improvement.
2. **Participant-Driven Reflection:** Participants share their perspectives, often revealing nuances others may have missed.
3. **Real-Time Feedback:** By debriefing immediately, participants retain vivid memories of the scenario, fostering meaningful discussions.

Pros of Traditional Debriefing

Traditional debriefing fosters an environment of open communication. Imagine a facilitator asking, "How did you feel during that critical moment when the patient's blood pressure dropped?" Participants can immediately process their emotions and link them to their actions.

- **Personalized Interaction:** Facilitators adapt discussions to meet the group's needs, diving deeper into areas of confusion or highlighting exceptional performance.
- **Fosters Psychological Safety:** Participants feel supported, knowing the focus is on learning rather than judgment.
- **Immediate Integration of Feedback:** Participants leave the session with actionable insights they can apply in subsequent scenarios or real-world settings.

Cons of Traditional Debriefing

Despite its benefits, traditional debriefing has limitations. Imagine relying solely on memory to recall intricate team dynamics or subtle nonverbal cues—details often essential for meaningful learning.

- **Reliance on Memory:** Participants may unintentionally omit or misremember critical aspects of the scenario.
- **Facilitator Dependency:** The quality of the debrief hinges on the facilitator's expertise and ability to navigate discussions effectively.

- **Limited Objectivity:** Without visual evidence, discussions may become subjective, potentially leading to biased conclusions.

Video Debriefing

Video debriefing incorporates recorded simulation footage into the debriefing process, allowing participants to review their actions and interactions objectively. It's as if participants become spectators of their performance, gaining a new perspective on their actions and decisions.

Key Characteristics

1. **Playback of Recorded Scenarios:** Video recordings allow participants to revisit specific moments, offering a concrete reference point.
2. **Objective Analysis:** Facilitators use video evidence to highlight key behaviors and events, eliminating reliance on memory alone.
3. **Enhanced Self-Reflection:** Participants observe their own body language, tone of voice, and decision-making, often leading to deeper insights.

Pros of Video Debriefing

Video debriefing unlocks opportunities for detailed analysis. Imagine a participant watching themselves in the video and exclaiming, "I didn't realize I interrupted my teammate during the discussion!" This realization may not surface in traditional debriefing.

- **Objective Feedback:** Provides a clear, unbiased account of events, enhancing accuracy and learning.
- **Enhanced Learning:** Participants can identify non-verbal cues, team dynamics, and overlooked details, fostering deeper understanding.
- **Reusability:** Videos can be revisited for future training or shared with other teams for benchmarking.
- **Promotes Accountability:** Seeing their actions fosters self-awareness and a sense of responsibility among participants.

Cons of Video Debriefing

However, video debriefing is not without challenges. For some participants, the thought of being recorded can be intimidating.

- **Technical Challenges:** Recording, playback, and storage require infrastructure and expertise, potentially increasing costs.

- **Time-Intensive:** Reviewing video footage adds to the debriefing duration, which may strain schedules.
- **Potential for Anxiety:** Participants may feel self-conscious, which could affect both their performance and engagement during debriefing.
- **Risk of Overemphasis:** Focusing excessively on minor details in the video may detract from broader learning objectives.

Comparing Traditional and Video Debriefing

Aspect	Traditional Debriefing	Video Debriefing
Feedback Source	Participant recollection and facilitator input	Objective video evidence and facilitator input
Reflection Method	Verbal discussion	Visual and auditory review with discussion
Facilitator Role	Central to guiding and interpreting events	Central but supported by video evidence
Time Efficiency	Generally shorter	Potentially longer due to video playback
Participant Experience	Encourages open discussion	Encourages self-awareness but may increase anxiety
Learning Depth	Relies on memory and subjective accounts	Enhances objectivity and self-reflection

The Value of Video Debriefing

Video debriefing offers unique advantages, particularly in scenarios requiring in-depth analysis of complex interactions, such as teamwork, communication, or crisis management. Imagine a simulation of an emergency cesarean delivery where split-second decisions and seamless teamwork are critical. Reviewing the video enables participants to scrutinize their actions in granular detail, uncovering insights that might otherwise be missed.

- **Improved Accuracy:** Facilitators and participants can identify specific actions and their impacts.
- **Enhanced Team Dynamics:** Teams can observe their interactions, fostering better collaboration and communication.
- **Standardized Training:** Recorded sessions can establish benchmarks or train future facilitators.

However, video debriefing should complement, not replace, traditional methods. For example, starting with a traditional discussion to address emotions before transitioning to video analysis can create a balanced and effective approach.

Challenges in Implementing Video Debriefing

Implementing video debriefing comes with its own set of challenges. For instance, some institutions may struggle to justify the initial investment in equipment, while others may face resistance from participants wary of being recorded.

- **Equipment and Cost:** Institutions must invest in cameras, software, and storage solutions, which can be prohibitive for smaller programs.
- **Facilitator Training:** Facilitators need skills to integrate video evidence effectively into discussions.
- **Privacy Concerns:** Clear video usage and confidentiality policies are essential to ensure participant trust and compliance.

Best Practices for Combining Approaches

- **Set Clear Objectives:** Define the purpose of using video in the debriefing process.
- **Blend Methods:** Use traditional debriefing to address emotions and initial reflections, followed by video analysis for deeper insights.
- **Focus on Key Moments:** Highlight specific video events aligned with learning objectives rather than reviewing the entire recording.
- **Ensure Psychological Safety:** Reassure participants about the purpose of recording and maintain confidentiality.
- **Train Facilitators:** Provide professional development to ensure facilitators can effectively manage traditional and video debriefing techniques.

Key Takeaways

1. Traditional debriefing emphasizes facilitated discussion and participant reflection, while video debriefing adds objectivity and depth through recorded evidence.

2. Video debriefing enhances learning by providing visual and auditory feedback but requires careful implementation to address logistical and emotional challenges.
3. Combining traditional and video debriefing can maximize learning outcomes, leveraging the strengths of both approaches.

Reflection

How can you integrate video debriefing into your current practices? Reflect on the balance between fostering psychological safety and leveraging the power of video to enhance reflective learning.

References

Cheng, A., Grant, V., Dieckmann, P., Arora, S., Robinson, T., & Eppich, W. (2016). Faculty development for simulation programs: Five issues for the future of debriefing training. Simulation in Healthcare, 11(4), 209-215.

Fanning, R. M., & Gaba, D. M. (2007). The role of debriefing in simulation-based learning. Simulation in Healthcare, 2(2), 115-125.

Rudolph, J. W., Simon, R., Rivard, P., Dufresne, R. L., & Raemer, D. B. (2007). Debriefing with good judgment: Combining rigorous feedback with genuine inquiry. Anesthesia & Analgesia, 104(3), 488-500.

Chapter 11
Prebrief: Understanding the Difference Between Military and Civilian Approaches

Introduction

We can't talk about debriefing and not include prebriefing. Prebriefing is critical to simulation-based learning, allowing participants to engage meaningfully with the scenario. It is akin to setting the tone before a theatrical performance, where each individual understands their role, the plot, and the desired outcomes. In healthcare simulation, prebriefing provides context, clarifies objectives, and prepares participants for the experiential learning process. The prebrief also serves as a psychological contract, ensuring participants understand expectations and the nonjudgmental nature of the exercise.

While prebriefing is universal, its application can vary significantly between military and civilian contexts. These differences stem from distinct organizational cultures, training goals, and operational realities. By diving into the intricacies of each approach, we can better appreciate how the prebrief lays the foundation for effective learning and performance in diverse settings.

The Purpose of Prebriefing

Imagine stepping into an unfamiliar environment with no clear understanding of what is expected. Prebriefing eliminates this uncertainty by establishing clarity and fostering readiness. The primary goals of prebriefing are:

- **Orientation:** Introducing participants to the simulation environment, tools, and scenario context.
- **Psychological Safety:** Creating an environment where participants feel safe to make mistakes and learn.
- **Expectations:** Clearly defining objectives, roles, and performance metrics.
- **Preparation:** Allowing participants to align their mental models with the scenario's intended learning outcomes.

The prebrief provides a map for the journey participants will undertake, whether in a military or civilian setting. However, the terrain of that journey differs significantly between the two domains.

Military Prebriefing

Picture a military prebrief as the start of a mission in a war movie—focused, intense, and deliberate. In military training, prebriefing is often highly structured and aligned with operational doctrine. The primary objectives include mission readiness, standard operating procedure (SOP) adherence, and rapid decision-making under stress. Key characteristics of military prebriefs include:

1. **Mission-Centric Approach:** Military prebriefs emphasize mission objectives and expected outcomes. This mirrors real-world operational demands where tasks are often high-stakes and time-sensitive. Participants are reminded that their training directly impacts lives and national security.
2. **Detailed Scenario Breakdown:** Participants are provided extensive details, including mission objectives, anticipated challenges, and contingencies. The "five-paragraph order" (situation, mission, execution, sustainment, command, and signal) is a common framework that ensures no stone is left unturned.
3. **Focus on Hierarchy and Command:** The military prebrief reinforces chain-of-command structures. Participants are briefed on roles and responsibilities, ensuring decision-making and operational flow clarity.

4. **Stress Inoculation:** Prebriefs often include elements designed to simulate the stress of real-world operations, preparing participants to maintain composure and effectiveness under pressure. For instance, a prebrief may outline how to respond when communication systems fail or unexpected adversaries emerge.

Example of a Military Prebrief

Scenario: Urban Combat Search and Rescue Simulation

Setting: The training facility is modeled after an urban environment with simulated hostile and civilian presence.

Facilitator Opening Statement: "Good morning, team. Today, we will conduct a search-and-rescue mission in a simulated urban environment. The objective is to locate and extract a high-value individual while navigating potential hostilities. Remember, this simulation is designed to mirror the complexities and stressors of real-world operations. Use this opportunity to apply your training, refine your skills, and learn from the experience."

Five-Paragraph Order:

Situation:

- Friendly forces are positioned in a secure location two clicks south of the mission site.
- Hostile presence is suspected in the northern quadrant, with reports of potential IEDs.
- Civilians may be present; exercise caution to minimize collateral damage.

Mission:

- Locate and extract the high-value individual (HVI) within one hour.
- Secure evidence of hostile activity if found.

Execution:

- The team will split into two units: Alpha for reconnaissance and Bravo for extraction.

- Coordinate movements using secure comms. Engage hostiles only when necessary.
- Report findings to the command at 15-minute intervals.

Sustainment:

- Medics are on standby at the secure location.
- Supplies, including ammunition and medical kits, are distributed evenly.

Command and Signal:

Call signs will be Alpha-One and Bravo-One.
Use Channel 5 for primary communications and Channel 3 for emergency signals.

Facilitator Closing Statement: "This mission will test your ability to adapt under pressure, communicate effectively, and adhere to the chain of command. Remember to prioritize the safety of your team and civilians. Debriefing will follow the exercise to identify strengths and areas for improvement. Execute with precision and stay mission-focused."

Civilian Prebriefing

Imagine a civilian prebrief as a team huddle before a complex surgical procedure. In civilian healthcare simulation, the emphasis is on individual and team learning in a supportive environment. Here, the facilitator is a guide, ensuring that participants feel empowered to explore, experiment, and grow. Key characteristics include:

1. **Learning-Centric Approach:** The focus is on education and skill development, rather than achieving specific mission objectives. Participants are encouraged to ask questions and think critically about their roles in the simulation.
2. **Psychological Safety:** Significant time is dedicated to creating a safe environment for participants. Facilitators emphasize that the simulation is a space for learning and growth, not judgment or punishment. For example, participants may be told, "Mistakes here are opportunities to learn."
3. **Flexible Scenarios:** Civilian prebriefs often allow for flexibility in participant actions, enabling exploration of various approaches to problem-solving. Unlike the rigid structure of military scenarios, civilian simulations may evolve based on participants' decisions.

4. **Emphasis on Teamwork:** The prebrief highlights the importance of interprofessional collaboration, reflecting the diverse nature of healthcare teams. Participants are encouraged to communicate openly and rely on each other's expertise.

Pre-Conferences

The pre-conference is a critical component of the prebrief process, ensuring that students are adequately prepared before engaging in their clinical or simulation experiences. Pre-conferences are used more so in nursing, respiratory therapy, medical assisting, EMTs/Paramedic clinical rotations where the students interact with a preceptor or instructor. This structured discussion allows educators to assess student readiness, clarify expectations, and reinforce key learning objectives.

Validating Student Assignment Preparation

It is essential to confirm students' preparedness before they begin their assigned activities. Educators can use the pre-conference to review completed pre-work, discuss assigned readings, and assess comprehension of foundational concepts. This process ensures that students arrive informed and ready to participate actively in the learning experience.

Ensuring Learner Readiness

Beyond academic preparation, the pre-conference helps determine whether students are mentally and emotionally prepared to enter their assigned clinical or simulation environment. This step allows instructors to identify any concerns, knowledge gaps, or anxieties affecting performance, enabling them to provide necessary support. Beyond the mental and emotional components, pre-conferences can be used to ensure learners are ready to represent the school or college they belong to by ensuring proper dress and appearance.

Reviewing Daily Expectations

Clearly outlining daily expectations helps students understand their roles and responsibilities. Educators can review specific tasks, procedural steps, and behavioral expectations during the pre-conference, ensuring alignment with learning objectives. Logistics, like when, where, and who to report to on the clinical location, are also relayed. This transparency fosters accountability and enhances student confidence.

Helping Set Priorities

Clinical and simulation experiences can be overwhelming, particularly for students still developing their decision-making skills. The pre-conference provides an opportunity to guide students in prioritizing tasks, managing time effectively, and understanding the most critical aspects of patient care or simulated scenarios.

Linking Theory to Clinical Practice

One of the primary goals of the pre-conference is to bridge the gap between theoretical knowledge and real-world application. Instructors can facilitate discussions that connect classroom learning with practical experiences, reinforcing key concepts and demonstrating their relevance in a clinical or simulated setting. By drawing these connections, students develop a deeper understanding of how theory informs clinical judgment and decision-making.

By incorporating these elements into the pre-conference, educators can enhance student engagement, improve learning outcomes, and create a more structured and supportive learning environment.

Pre-Conference
• Validate Student Assignment Preparation • Ensure Learner Readiness • Review of Daily Expectations • Help Set Priorities • Link Theory to Clinical Practice

Setting Up a Healthcare Simulation Prebrief Using Best Practices

Creating an effective healthcare simulation prebrief requires careful planning and adherence to best practices. Below are the essential steps and elements to include:

Steps to Set Up the Prebrief

1. **Define Objectives:** Clearly outline the learning goals for the simulation. These objectives should align with the overall curriculum and address specific skills, knowledge, or behaviors.
2. **Prepare the Environment:** Ensure the simulation space is organized and has all equipment and tools in place. A brief walkthrough of the physical environment can help participants feel more comfortable.
3. **Establish Psychological Safety:** Begin by setting a positive tone. Emphasize that the simulation is a safe space for learning and that mistakes are expected and valuable for growth.
4. **Outline Roles and Expectations:** Clearly define the roles of participants, facilitators, and any observers. Concisely explain what is expected during the scenario and the debriefing process.
5. **Provide Context:** Offer a narrative that sets the scene for the simulation. Include background information, such as the patient's history or the situational context, to help participants mentally prepare.
6. **Discuss Logistics:** Explain the simulation flow, including how time will be managed, when interventions may occur, and how feedback will be provided.

Essential Elements to Include

- **Scenario Overview:** Briefly explain the clinical situation, goals, and specific challenges participants may encounter.
- **Confidentiality Agreement:** A reminder that the simulation and discussions are confidential and intended solely for learning purposes.
- **Ground Rules:** Establish rules for interaction, such as respectful communication and staying in role during the scenario.
- **Technical Orientation:** Demonstrate any unfamiliar equipment or technology to avoid confusion during the simulation.
- **Encouragement of Questions:** Allow participants to clarify any uncertainties before beginning.

By incorporating these steps and elements, facilitators can create a prebrief that prepares participants for the simulation and fosters a productive and supportive learning environment.

Example of a Civilian Prebrief

Scenario: Emergency Room Trauma Simulation

Setting: Simulation lab designed to replicate an emergency department, complete with patient mannequins, monitors, and medical equipment.

Facilitator Opening Statement: "Welcome, everyone. Today, we are simulating a high-acuity trauma case in the emergency department. Our goal is to work collaboratively as an interprofessional team to stabilize the patient and prioritize their care. Remember, this is a safe space for learning—there are no wrong answers, only opportunities to improve."

Key Elements of the Prebrief:

Scenario Overview:
"The patient is a 35-year-old male involved in a motor vehicle collision. He is arriving with multiple injuries and signs of shock. Your task is to assess, stabilize, and prepare the patient for surgery."

Team Roles and Responsibilities:
"Each of you will play a specific role: primary nurse, physician, respiratory therapist, and scribe. Please refer to the role cards provided."

Psychological Safety:
"This simulation is about learning and improving. Feel free to ask questions and try different approaches. Remember, mistakes here are valuable learning tools."

Technical Orientation:
"Let's review the equipment you may need. The crash cart is located here, and the ventilator settings are pre-configured for this scenario. Please let me know if you have any questions."

Ground Rules:
"Stay in character as much as possible to maintain realism. If you need to step out, raise your hand and say 'pause.'"

Facilitator Closing Statement: "Let's focus on communication, teamwork, and clinical decision-making. After the simulation, we will debrief to reflect on what went well and identify areas for growth. Let's begin!"

Example of a Civilian Pre-Surgical Prebrief

Location: Operating Room (OR) Conference Area

Participants: Surgeon, Anesthesiologist, Circulating Nurse, Scrub Nurse, Surgical Technician, and OR Manager

Facilitator: Lead Surgeon

Prebrief Structure

1. Introduction (Facilitator-Led)
Purpose: "This prebrief is to ensure clarity on the plan for today's procedure and align the team on patient safety, surgical protocols, and contingency measures."

Emphasis on Communication: "Please speak up if you notice anything that could compromise the safety or success of this procedure. Every perspective is valuable."

2. Patient Overview

Patient Information:
Name: John Smith, 55 years old
Procedure: Right Total Knee Replacement
Relevant History: Hypertension, mild sleep apnea, and prior reaction to anesthesia.
Circulating Nurse shares patient chart highlights.

3. Surgical Plan Review

Lead Surgeon:
"We're performing a standard right total knee replacement. The expected duration is 90 minutes. I'll guide the surgical steps, but the main considerations are addressing osteophytes in the posterior knee and verifying alignment for optimal recovery."

Anesthesiologist:
"Patient will be under general anesthesia. Airway monitoring is critical due to sleep apnea. We'll use additional precautions, including continuous capnography."

Roles and Responsibilities
Surgeon: Oversees procedure and critical decision-making.
Scrub Nurse: Prepares and passes instruments.
Circulating Nurse: Coordinates supplies and documents intraoperative events.
Anesthesiologist: Monitors vitals and manages airway.
Surgical Tech: Assists with equipment and maintains sterile field.

5. Anticipated Challenges and Mitigation
Lead Surgeon: "Due to the patient's medical history, increased bleeding is possible. Ensure suction equipment is ready, and have blood products on standby."
Anesthesiologist: "Let's watch for any airway difficulties and maintain heightened awareness during induction and emergence."

6. Contingency Plan
Emergency protocols reviewed:
Massive Bleeding: Notify blood bank immediately.
Anesthesia Reaction: Epinephrine and crash cart will be ready.

7. Closing and Final Questions
Facilitator: "Does anyone have questions or concerns? Are there additional issues we need to address?"

Team confirms readiness.

Key Differences Between Military and Civilian Prebriefing

Aspect	Military	Civilian
Purpose	Mission readiness and SOP adherence	Skill development and reflective learning
Structure	Highly structured and doctrine-driven	Flexible and learner-centered

Focus	Chain of command and operational outcomes	Teamwork and individual growth
Psychological Safety	Implicit, with an emphasis on performance	Explicit, with an emphasis on learning
Stress Environment	Designed to simulate operational stress	Supportive and low-pressure

Challenges and Best Practices

Challenges

- **Cultural Differences:** Adapting military-style prebriefs to civilian settings can lead to misunderstandings, as civilian learners may be less familiar with hierarchical structures. Conversely, civilian methods in military contexts may undermine the precision and discipline required for operational success.
- **Objective Alignment:** Balancing the mission-centric focus of military prebriefs with the learning-centered goals of civilian simulations can be challenging.
- **Psychological Safety:** Ensuring participants feel safe making mistakes is emphasized more in civilian contexts but remains critical in both domains. Achieving this balance requires skillful facilitation.

Best Practices

- **Tailor the Approach:** Design prebriefs to align with the participants' organizational culture and learning objectives. For example, facilitators can blend structured and flexible elements in mixed civilian-military simulations.
- **Foster Psychological Safety:** Regardless of the setting, emphasize the importance of a nonjudgmental environment. Facilitators should explicitly address this, especially in high-stakes scenarios.
- **Leverage Commonalities:** Focus on universal elements of prebriefing, such as clarity of objectives and participant preparation, to create a shared understanding.

Key Takeaways

- Prebriefing is essential in military and civilian simulations but differs in structure, purpose, and focus.

- Military prebriefs are mission-centric, structured, and stress-oriented, while civilian prebriefs prioritize learning, teamwork, and psychological safety.
- Understanding these differences can enhance the design and effectiveness of simulations across contexts, ensuring that participants are adequately prepared and engaged.

Reflection

Consider how the principles of prebriefing can be adapted to bridge gaps between military and civilian simulation practices. Reflect on your own experiences: How can facilitators create environments that balance the rigor of military training with the flexibility and psychological safety emphasized in civilian settings? How might you design prebriefs that honor both approaches?

References

Cheng, A., Palaganas, J., Eppich, W., Rudolph, J., Robinson, T., & Grant, V. (2016). Co-debriefing for simulation-based education: A primer for facilitators. *Simulation in Healthcare, 11*(1), 10-15.

Rudolph, J. W., Raemer, D. B., & Simon, R. (2014). Establishing a safe container for learning in simulation: The role of the pre-simulation briefing. *Simulation in Healthcare, 9*(6), 339-349.

Salas, E., Tannenbaum, S. I., Kraiger, K., & Smith-Jentsch, K. A. (2012). The science of training and development in organizations: What matters in practice. *Psychological Science in the Public Interest, 13*(2), 74-101.

U.S. Army Field Manual (FM) 6-0. (2014). Commander and staff organization and operations. Washington, DC: Department of the Army.

Weller, J. M., Nestel, D., Marshall, S. D., Brooks, P. M., & Conn, J. J. (2012). Simulation in clinical teaching and learning. *Medical Journal of Australia, 196*(9), 594-594.

Chapter 12
Debriefing is Not Easy

Debriefing is a cornerstone of effective healthcare education and team performance improvement. However, it is not as straightforward as it might appear. Facilitating a successful debrief requires practice, skill, and an understanding of both the art and science of the process. This chapter explores the challenges associated with debriefing and emphasizes why it is a skill that demands ongoing refinement.

The Complexity of Debriefing

At its core, debriefing is a structured conversation that allows teams to reflect on actions, decisions, and outcomes. While the concept may sound simple, the reality is far more intricate. Effective debriefing involves balancing psychological safety, managing emotions, and focusing on learning objectives. These elements do not naturally come together without deliberate effort and preparation.

One of the primary challenges is fostering an environment of psychological safety. Participants must feel comfortable sharing their thoughts, admitting mistakes, and engaging in honest dialogue. This can be difficult, particularly in hierarchical environments or when individuals fear judgment or criticism. Facilitators must work intentionally to create a safe and supportive space, which takes time and sensitivity.

Another challenge is navigating the emotional dynamics of a debrief. Clinical events, particularly high-stakes or emotionally charged ones, can leave participants feeling vulnerable, defensive, or even traumatized.

Facilitators must recognize and address these emotions while steering the conversation toward constructive reflection and learning. This requires emotional intelligence, empathy, and the ability to adapt in real time.

The Skills Required for Effective Debriefing

Successful debriefing is a learned skill, and even experienced facilitators must continuously refine their techniques. Some of the key competencies include:

1. Active Listening

Facilitators must actively listen to participants, not just to their words but also to the underlying emotions and concerns. This requires full attention, nonverbal cues, and asking clarifying questions to ensure understanding. Active listening helps participants feel heard and valued, which is critical for fostering engagement.

2. Managing Group Dynamics

Groups often include diverse personalities, perspectives, and levels of experience. Facilitators must manage these dynamics to ensure that all voices are heard and that dominant participants do not overshadow others. This involves encouraging quieter team members to share their thoughts and gently redirecting discussions that veer off track.

3. Asking Effective Questions

The quality of the debrief is often determined by the quality of the questions. Facilitators must craft open-ended, thought-provoking questions that prompt reflection and analysis. For example, instead of asking, "Why didn't you follow the protocol?" a facilitator might ask, "What factors influenced your decision-making in that moment?"

4. Providing Constructive Feedback

Giving feedback is one of the most challenging aspects of debriefing. Facilitators must strike a balance between highlighting areas for improvement and recognizing strengths. Feedback should be specific, actionable, and delivered in a way that motivates rather than discourages participants.

5. Time Management

Debriefs often occur in time-constrained environments, yet they must be thorough enough to extract meaningful lessons. Facilitators must prioritize key issues, keep discussions focused, and ensure all critical points are addressed within the allotted time.

Common Challenges in Debriefing

Despite preparation and skill, facilitators often encounter challenges that can derail a debrief. Some common obstacles include:

1. Resistance to Participation

Participants may be reluctant to engage for various reasons, such as fear of criticism, fatigue, or a lack of understanding about the purpose of the debrief. Overcoming this resistance requires patience, reassurance, and clear communication about the process's value.

2. Emotional Responses

Emotional outbursts or defensiveness can arise during debriefs, particularly after adverse events. Facilitators must address these emotions empathetically while redirecting the conversation toward constructive analysis.

3. Lack of Focus

Debriefs can sometimes devolve into tangential discussions or complaints, which dilute their effectiveness. Facilitators must gently steer the conversation to the learning objectives without dismissing participants' concerns.

4. Hierarchical Barriers

In hierarchical teams, junior members may hesitate to speak openly, particularly if senior staff are present. Facilitators must actively work to level the playing field by encouraging equal participation and modeling respectful dialogue.

The Role of Practice in Mastery

Becoming proficient in debriefing is not an overnight process. It requires practice, self-reflection, and a willingness to learn from mistakes. New facilitators may feel overwhelmed or uncertain at first, but with experience, their confidence and competence grow.

One effective way to build debriefing skills is through simulation. Facilitators can practice leading debriefs in a controlled environment, receive feedback from peers, and refine their techniques. Role-playing different scenarios—such as managing a defensive participant or addressing a sensitive topic—can also help facilitators prepare for real-world challenges.

Following a structured framework, such as a debriefing script or prompts, can make the process more manageable, especially for new facilitators. Scripts provide a roadmap for navigating complex conversations and ensure that key learning objectives are addressed.

Another valuable strategy is co-facilitating with a more experienced debriefer. Watching a skilled facilitator in action can provide new insights and build confidence. Co-facilitation also allows for real-time feedback and shared responsibility, making the process less intimidating for novice debriefers.

The more facilitators practice, the better they become. Repeated exposure to different scenarios helps refine skills and builds familiarity with common challenges. Recording debriefs for process improvement can also be highly effective. Reviewing recordings allows facilitators to identify areas for improvement and adjust their approach accordingly.

Consulting and learning from experts in healthcare simulation or debriefing techniques is another pathway to growth. Experts can provide mentorship, share best practices, and offer constructive feedback that accelerates skill development.

Key Takeaways

- **Debriefing is a skill:** Effective debriefing requires practice, preparation, and continuous improvement. It is not an innate talent but a competency that can be developed over time.
- **Psychological safety is paramount:** Creating a safe and supportive environment is essential for fostering honest dialogue and meaningful reflection.
- **Facilitators need a range of skills:** From active listening to managing emotions, successful debriefing involves a combination of interpersonal and technical competencies.
- **Challenges are inevitable:** Resistance, emotions, and hierarchical dynamics can complicate debriefs, but skilled facilitators learn to navigate these obstacles.
- **Practice makes progress:** Simulation, feedback, and professional development are critical for building confidence and competence in debriefing.
- **Frameworks and mentorship help:** Using scripts, co-facilitating with experienced debriefers, and consulting experts can significantly enhance debriefing effectiveness.

Reflection

Reflect on your own experiences with debriefing—whether as a participant or a facilitator. What challenges have you encountered, and how did you address them? What steps can you take to improve your debriefing skills or foster a more supportive environment for others?

By recognizing that debriefing is not easy and committing to ongoing practice, healthcare professionals can unlock the full potential of this powerful tool for learning and improvement.

References

Cheng, A., Grant, V., Dieckmann, P., Arora, S., Robinson, T., & Eppich, W. (2016). Debriefing for technology-enhanced simulation: A systematic review and meta-analysis. *Medical Education, 50*(4), 469-478. https://doi.org/10.1111/medu.12975

Dieckmann, P., Gaba, D., & Rall, M. (2009). Deepening the theoretical foundations of patient simulation as social practice. *Simulation in Healthcare, 4*(3), 183-193. https://doi.org/10.1097/SIH.0b013e3181a2dd0d

Edmondson, A. C. (2019). *The fearless organization: Creating psychological safety in the workplace for learning, innovation, and growth.* Wiley.

Eppich, W., & Cheng, A. (2015). Promoting excellence and reflective learning in simulation (PEARLS). *Simulation in Healthcare, 10*(2), 106-115. https://doi.org/10.1097/SIH.0000000000000072

Fanning, R. M., & Gaba, D. M. (2007). The role of debriefing in simulation-based learning. *Simulation in Healthcare, 2*(2), 115-125. https://doi.org/10.1097/SIH.0b013e3180315539

Kolbe, M., Grande, B., & Spahn, D. R. (2013). The ABC of debriefing: A systematic approach to feedback in simulation. *Advances in Simulation, 2*(1), 2-12. https://doi.org/10.1186/s41077-016-0023-1

Raemer, D. B., Anderson, M., Cheng, A., Fanning, R. M., Nadkarni, V., & Savoldelli, G. L. (2011). Research regarding debriefing as part of the learning process. *Simulation in Healthcare, 6*(7), S52-S57. https://doi.org/10.1097/SIH.0b013e318229f5eb

Rudolph, J. W., Simon, R., Dufresne, R. L., & Raemer, D. B. (2006). There's no such thing as "nonjudgmental" debriefing: A theory and method for debriefing with good judgment. *Simulation in Healthcare, 1*(1), 49-55. https://doi.org/10.1097/01266021-200600110-00006

Salas, E., Klein, C., King, H., Salisbury, M., Augenstein, J. S., Birnbach, D. J., & Fernandez, P. (2008). Debriefing medical teams: 12 evidence-based best practices and tips. *Joint Commission Journal on Quality and Patient Safety, 34*(9), 518-527. https://doi.org/10.1016/S1553-7250(08)34066-5

Sawyer, T., Eppich, W., Brett-Fleegler, M., Grant, V., & Cheng, A. (2016). More than one way to debrief: A critical review of healthcare simulation debriefing methods. *Simulation in Healthcare, 11*(3), 209-217. https://doi.org/10.1097/SIH.0000000000000148

Chapter 13
Tools for Debriefing and Tips for Best Practice

Introduction

Effective debriefing is a skill that combines the right tools with thoughtful facilitation to maximize learning. From structured frameworks to advanced technological aids, the tools available for debriefing have evolved significantly. Imagine a team of healthcare professionals gathered after a high-pressure simulation. The facilitator initiates a discussion using prompts like, "What went well?" and "What could we improve?" Now consider how their learning experience could be enriched by adding video playback, checklists, or virtual reality tools to the mix. Whether conducting a traditional discussion-based debrief or leveraging video playback, selecting the appropriate tools and applying best practices can elevate the debriefing process.

This chapter explores the various tools used in debriefing, their applications, and practical tips to ensure successful implementation.

Tools for Debriefing

1. Structured Frameworks

Frameworks provide a roadmap for debriefing sessions, ensuring that discussions are organized, goal-oriented, and comprehensive. Commonly used frameworks include:

- **GAS Model (Gather, Analyze, Summarize):** Focuses on collecting participant input, analyzing actions, and summarizing lessons. For example, a facilitator might gather reactions by asking, "What were your initial thoughts?" then guide the team to analyze a critical decision point before summarizing lessons learned.
- **PEARLS (Promoting Excellence and Reflective Learning in Simulation):** Combines open-ended inquiry with structured reflection, encouraging participants to think critically about their actions.
- **Diamond Model:** Begins with a broad discussion to allow emotional processing, narrows focus during analysis and concludes with actionable takeaways.

When to Use: These frameworks guide discussions in traditional or video debriefing sessions. Facilitators can adapt them based on participant needs and learning objectives.

2. Video Playback Tools

Video tools allow participants to review recorded simulations, enhancing objectivity and reflection. For instance, a team might replay a moment where communication broke down during a high-pressure scenario, discussing how clearer messaging could have improved outcomes. Popular tools include:

- **Standard Video Recorders:** Simple cameras or smartphones can capture simulations for playback.
- **Integrated Simulation Platforms:** Systems like Laerdal's SimCapture or CAE LearningSpace offer synchronized video recording, annotation, and playback features.

When to Use: Video playback is particularly useful for analyzing team dynamics, communication, and technical performance in detail. Watching themselves in action often gives participants eye-opening insights into their behaviors and interactions.

3. Facilitation Guides and Checklists

These tools help facilitators maintain structure and consistency during debriefing. Examples include:

- **Debriefing Guides:** Provide prompts and questions aligned with learning objectives. For example, "How did you prioritize your tasks when the patient's condition worsened?"
- **Behavioral Checklists:** Outline specific actions or skills to address, such as effective communication or protocol adherence. These tools ensure that no critical element is overlooked.

When to Use: Use guides and checklists to address key learning points, especially when debriefing large groups or complex scenarios.

4. Technology-Enhanced Tools

Advanced technologies are transforming debriefing, offering interactive and immersive experiences. Examples include:

- **Virtual Reality (VR) Platforms:** VR simulations can be paused and replayed during debriefing, allowing participants to analyze scenarios from multiple perspectives. For example, participants could "walk through" a simulated emergency room to understand spatial dynamics.
- **Debriefing Software:** Tools like iSimulate and B-Line Medical enable facilitators to tag key moments in real-time for later discussion.

When to Use: These tools are ideal for high-fidelity simulations and scenarios requiring advanced analysis. They provide a level of interactivity that deeply engages participants.

5. Feedback and Evaluation Tools

- **Self-Assessment Forms:** Encourage participants to reflect on their performance before the debriefing. For example, participants might rate their teamwork or communication skills on a scale of 1-5.
- **Facilitator Feedback Forms:** Provide structured feedback to participants, focusing on strengths and areas for improvement.

When to Use: These tools are valuable for fostering self-awareness and creating actionable learning plans. They help participants take ownership of their development.

<u>Self-Assessment Forms</u>

Self-assessment forms are tools designed to help participants reflect on their performance during a simulation. These forms often include structured questions and rating scales that prompt participants to evaluate their skills, decision-making, and teamwork.

Purpose and Rationale

- Encourage participants to take ownership of their learning.
- Provide insight into participants' self-perception and areas of confidence or concern.
- Foster critical thinking and self-awareness by prompting reflection on actions and decisions.

Type of Data Provided

- **Quantitative Data:** Rating scales (e.g., 1-5) on specific competencies like communication, leadership, or technical skills.
- **Qualitative Data:** Open-ended responses describing challenges faced, areas for improvement, and lessons learned.

When to Use: Self-assessment forms are most effective when completed immediately after the simulation but before the debriefing session. This ensures that participants' reflections are fresh and can contribute meaningfully to the discussion.

Example Self-Assessment Form

Category	Question/Prompt	Rating (1-5)
Communication	How effectively did I communicate with my team during the scenario?	
Decision-Making	How confident was I in my clinical decision-making under pressure?	
Teamwork	How well did I collaborate with other team members to achieve the scenario's goals?	
Technical Skills	How proficient was I in performing technical tasks required during the simulation?	
Overall Reflection	What did I do well, and what could I improve next time?	**Open-Ended**

Using the Data: Facilitators can use participants' self-assessment data to:

- Identify trends or common challenges among the group.
- Guide the debriefing session to address specific areas of concern.
- Provide individualized feedback that aligns with participants' self-perceptions.

Facilitator Feedback Forms

Facilitator feedback forms are structured tools that allow facilitators to provide targeted, actionable feedback to participants after a simulation. They are designed to evaluate specific behaviors, skills, and decision-making processes, ensuring that feedback is constructive and tailored to individual needs.

Purpose and Rationale

- Provide participants with clear and objective evaluations of their performance.
- Highlight strengths and identify specific areas for improvement.
- Encourage accountability and professional growth through actionable feedback.
- Support consistency across debriefing sessions by standardizing feedback delivery.

Type of Data Provided

- **Behavioral Observations:** Facilitators document specific actions or behaviors exhibited during the simulation.
- **Skill-Based Assessments:** Ratings or evaluations of clinical, communication, and teamwork competencies.
- **Developmental Feedback:** Suggestions for improvement and strategies for future success.

When to Use: Facilitator feedback forms are most effective when completed immediately after the simulation and integrated into the debriefing session. They can also be shared with participants in a formal evaluation or learning plan.

Example Facilitator Feedback Form

Category	Observation/Feedback	Suggestions for Improvement

Communication	Demonstrated clear communication but occasionally interrupted others.	Practice active listening and allowing others to speak.
Decision-Making	Made timely decisions but hesitated in moments of uncertainty.	Develop confidence in high-stakes decision-making.
Teamwork	Worked well with the team but dominated discussions at times.	Encourage input from quieter team members.
Technical Skills	Performed procedures effectively but struggled with equipment setup.	Review and practice equipment setup protocols.
Overall Performance	Strong engagement and focus throughout the scenario.	Continue refining strengths and addressing key gaps.

Using the Data:

- Facilitators can use the feedback form to:
  - Provide structured feedback that aligns with simulation objectives.
  - Highlight specific examples to reinforce learning points.
  - Tailor recommendations to each participant, supporting their ongoing development.

Published Debriefing Tools

Several validated debriefing tools are widely used in simulation-based education. These tools have undergone rigorous testing for reliability and validity, ensuring their effectiveness in enhancing learning outcomes. Examples include:

- **Debriefing Assessment for Simulation in Healthcare (DASH):** Evaluates the quality of debriefing sessions using structured criteria. Facilitators can use DASH to assess their own performance and identify areas for improvement.
- **Objective Structured Assessment of Debriefing (OSAD):** Focuses on evaluating the debriefing process, particularly the facilitator's ability to engage participants and meet learning objectives.
- **TEAM Tool:** Assesses team performance during simulations, emphasizing communication, leadership, and collaboration.

Advantages of Published Tools:

- **Proven Effectiveness:** These tools are backed by research, ensuring they align with best practices and educational standards.
- **Standardization:** Provide a consistent framework for debriefing, ensuring uniformity across sessions.
- **Comprehensive Evaluation:** Address multiple dimensions of debriefing, including facilitator skills, participant engagement, and learning outcomes.
- **Adaptability:** Many tools can be customized to suit specific scenarios or institutional needs.

Using Published Tools:

- Integrate validated tools like DASH or OSAD into your debriefing practices to ensure a high standard of facilitation.
- Leverage the tools to provide structured, actionable feedback that aligns with institutional goals and participant needs.
- Use data from these tools to track improvements and refine debriefing practices.

Tips for Best Practice

1. Prepare Thoroughly

Familiarize yourself with the simulation's objectives and key learning points. For example, review video recordings or checklists in advance to identify critical moments for discussion.

Ensure all tools (e.g., video equipment, guides) are functioning and accessible. Technical issues can disrupt the session's flow and diminish its impact.

2. Create a Safe Learning Environment

Emphasize that debriefing is a nonjudgmental space for reflection and growth. For instance, start the session by saying, "This is a safe space to discuss what happened and how we can improve."

Encourage open communication by setting ground rules for respect and confidentiality.

3. Adapt to Participants' Needs

Tailor your approach based on the group's experience level, learning styles, and emotional responses. For example, novice participants may need more guidance, while experienced teams might benefit from more open-ended discussions.

Use open-ended questions to invite diverse perspectives. Questions like, "What was your reasoning behind that decision?" can spark rich discussions.

4. Leverage Visual and Audio Aids

Use video playback selectively, focusing on moments that align with learning objectives. For instance, highlight a key moment where communication faltered or teamwork excelled. Highlight non-verbal cues, team dynamics, and other aspects that may not be apparent in traditional debriefing.

5. Balance Structure and Flexibility

Follow a structured framework but remain adaptable to participants' contributions and unexpected discussion points. For example, if participants raise an unanticipated but valuable issue, allow room to explore it.

6. Engage All Participants

Ensure everyone has the opportunity to share their insights and reflections. Use techniques like round-robin discussions or small-group breakouts to involve quieter participants.

Use prompts like "What would you do differently next time?" to encourage self-reflection and improvement.

7. Summarize Key Takeaways

Conclude the session by summarizing key learning points and actionable steps. For example, "Today we learned the importance of clear communication during handoffs. Let's practice this in the next simulation."

Reinforce how the lessons can be applied in future scenarios or real-world practice.

Challenges and Solutions

Challenge 1: Limited Time for Debriefing

Solution: Use time-efficient frameworks like the GAS model and focus discussions on high-priority issues. For example, identify one or two critical learning points to explore in-depth.

Challenge 2: Technical Issues with Video Tools

Solution: Test equipment in advance and have backup tools available, such as written notes or guides. Being prepared ensures that technical glitches don't derail the session.

Challenge 3: Participant Resistance

Solution: Reassure participants about the purpose of debriefing and emphasize the value of constructive feedback. Highlight how the process supports their growth and improves team dynamics.

Key Takeaways

1. Selecting the right tools for debriefing enhances learning and reflection.
2. Combining validated frameworks with innovative technologies can deepen participants' understanding and engagement.
3. Published tools provide a reliable, standardized approach to ensure debriefing effectiveness and consistency.

Reflection

How can validated tools like DASH or OSAD enhance your debriefing sessions? Reflect on opportunities to integrate these tools into your practices for more impactful learning outcomes.

References

Cant, R. P., & Cooper, S. J. (2014). Simulation-based learning in nurse education: Systematic review. *Journal of Advanced Nursing, 66*(1), 3-15.

Cheng, A., Grant, V., Dieckmann, P., Arora, S., Robinson, T., & Eppich, W. (2016). Faculty development for simulation programs: Five issues for the future of debriefing training. *Simulation in Healthcare, 11*(4), 209-215.

Fanning, R. M., & Gaba, D. M. (2007). The role of debriefing in simulation-based learning. *Simulation in Healthcare, 2*(2), 115-125.

INACSL Standards Committee. (2016). INACSL standards of best practice: Simulation debriefing. *Clinical Simulation in Nursing, 12*(S), S21-S25. Brett-Fleegler, M., Rudolph, J., Eppich, W., Monuteaux, M., Fleegler, E., & Cheng, A. (2012). Debriefing assessment for simulation in healthcare: Development and psychometric properties. *Simulation in Healthcare, 7*(5), 288-294.

Jaye, P., Thomas, L., & Reedy, G. (2015). "The Diamond": A structure for simulation debrief. *The Clinical Teacher, 12*(3), 171-175.

Rudolph, J. W., Simon, R., Rivard, P., Dufresne, R. L., & Raemer, D. B. (2007). Debriefing with good judgment: Combining rigorous feedback with genuine inquiry. *Anesthesia & Analgesia, 104*(3), 488-500.

Chapter 14 Ethical Considerations for Debriefing in Healthcare Simulation

Debriefing in healthcare simulation is a powerful tool for learning, reflection, and improving patient safety. However, the process raises significant ethical considerations that must be carefully managed to ensure psychological safety, confidentiality, fairness, and respect for all participants. Ethical debriefing practices are essential to maintaining trust, fostering open communication, and protecting the well-being of healthcare professionals. This chapter explores the ethical dimensions of debriefing in healthcare simulation and offers guidance for navigating potential ethical challenges through real-world examples and best practices.

Psychological Safety and Emotional Well-Being

The Importance of Psychological Safety

Psychological safety is an environment where participants feel comfortable sharing their thoughts, experiences, and mistakes without fear of embarrassment, punishment, or retaliation. Ethical debriefing requires facilitators to foster a nonjudgmental space that prioritizes learning over blame.

Navigating Ethical Risks in Debriefing

Debriefing, by its nature, asks participants to confront their own decisions and actions, which can sometimes be an uncomfortable process. Participants may worry about how their words will be perceived or if their admissions of error will have negative consequences. Some common concerns include:

Fear of Judgment

Imagine being a junior nurse in a debriefing session with senior physicians. A scenario is presented where a critical medication was delayed, and the nurse knows they hesitated before administering it. The weight of eyes on them can be overwhelming, making it difficult to admit the truth. Without psychological safety, this hesitation can lead to silence, reducing the debriefing's effectiveness.

Emotional Distress

Some cases in healthcare carry heavy emotional weight. A resident in an emergency simulation dealing with a pediatric trauma patient might relive a real-life case that ended tragically. If not facilitated correctly, the debriefing session could reopen wounds rather than promote learning.

Power Imbalances

In many debriefings, participants range from senior physicians to interns and nurses. Junior staff may be reluctant to speak openly, especially if their experiences contradict senior colleagues' views. If power imbalances are not addressed, valuable insights can be lost.

Best Practices for Ethical Debriefing

- Establish **ground rules** for respectful discussion before the debriefing begins, ensuring a safe space for all participants.
- Encourage **constructive reflection** rather than punitive feedback, focusing on learning instead of blame.
- Normalize mistakes by framing them as opportunities for growth, emphasizing that errors are a universal part of the learning process.
- Provide **opt-out options** or access to mental health resources if a participant becomes distressed, reinforcing that emotional well-being is a priority.

Example

In a major teaching hospital, a simulation session on neonatal resuscitation was conducted. During the debriefing, a nurse hesitated to voice concerns about a colleague's decision. The facilitator, recognizing the nurse's hesitation, gently reiterated the purpose of the debriefing: to learn, not to assign blame. This reassurance created an open space where the nurse felt safe to speak. The resulting discussion led to clearer role definitions during crises, ultimately improving the team's future performance and fostering a culture of trust.

Confidentiality and Data Protection

Confidentiality in debriefing ensures that sensitive discussions remain private and do not negatively impact participants' reputations or careers. Healthcare professionals must feel assured that their words will not be used against them but rather as tools for learning and improvement.

The Challenge of Protecting Sensitive Information

Imagine a surgeon recounting a procedural error during a debriefing session. Without clear assurances of confidentiality, they might withhold key insights out of fear that their admission could lead to disciplinary action. Similarly, a nurse who hesitates before administering medication may worry that discussing the mistake openly could result in professional consequences rather than constructive feedback.

Potential Ethical Risks

- **Breach of Confidentiality:** If discussions are not kept private, participants may fear professional repercussions, leading to reduced honesty and engagement.
- **Unauthorized Use of Recordings:** Some debriefings involve video or audio recordings for review. Without strict regulations, these recordings could be misused, leading to breaches of trust.
- **Anonymity Concerns:** If debriefing data is used for research or performance evaluation, participants must be fully informed and assured that their identities will remain protected.

Best Practices for Ethical Confidentiality Management

- Clearly communicate that all debriefing discussions are **confidential** and will not be used for punitive purposes.
- Obtain **informed consent** before recording debriefing sessions and specify how recordings will be used and stored.
- De-identify participant data if used for research, ensuring that individuals cannot be traced back to specific comments or actions.

Example

A surgical department introduced a mandatory video-assisted debriefing process to improve post-operative outcomes. Initially, staff expressed concerns about privacy and potential misuse of recordings. The administration addressed these fears by establishing clear policies: recordings were only reviewed by facilitators, stored securely, and deleted after a set period. This reassured participants, leading to greater openness and engagement in the debriefing process.

Balancing Transparency and Ethical Sensitivity

The Need for Transparency in Debriefing

Effective debriefing relies on openness and honesty. Without transparency, participants may not fully engage or recognize areas for improvement. However, transparency must be handled with care to prevent harm.

The Risks of Excessive Transparency

Consider a scenario in which a physician makes a critical error during a simulation. If the debriefing focuses too heavily on this mistake without considering emotional sensitivity, the physician may feel embarrassed or demoralized, reducing the session's effectiveness. Similarly, if participation in a debriefing session is forced rather than voluntary, individuals may shut down rather than contribute meaningfully.

Potential Ethical Risks

- **Overexposure:** Highlighting an individual's error too prominently can lead to embarrassment or self-doubt, making it less likely they will engage in future debriefings.

- **Forced Participation:** Participants should feel free to engage at their own comfort level. If pressured, they may provide guarded responses rather than honest reflections.
- **Subjectivity and Bias:** Facilitators may unintentionally favor certain perspectives or focus on specific individuals rather than maintaining an unbiased discussion.

Best Practices for Ethical Transparency

- Use **objective language** and **neutral facilitation** techniques to ensure fairness.
- Allow participants to pass on questions or contribute at their comfort level.
- Focus discussions on **systems-based learning**, ensuring that the conversation promotes overall team improvement rather than individual blame.

Key Takeaways

- Psychological safety must be maintained to encourage open and constructive discussions without fear of judgment.
- Confidentiality is crucial to protecting participants and fostering a culture of trust in simulation debriefing.
- Transparency should be balanced with ethical sensitivity, ensuring discussions focus on systemic improvements rather than personal blame.
- Facilitators play a critical role in navigating emotional dynamics, avoiding bias, and ensuring fair and inclusive participation.
- Ethical use of technology, such as video recording and AI-driven analytics, requires careful consideration of privacy laws and participant consent.
- Organizations must create policies that promote ethical debriefing practices while providing training and resources for facilitators.

Reflection

How can you integrate ethical considerations into your debriefing practices? What steps can your organization take to ensure debriefing remains a psychologically safe and constructive learning tool?

References

Edmondson, A. C. (2019). *The Fearless Organization: Creating Psychological Safety in the Workplace for Learning, Innovation, and Growth.* Wiley.

Fanning, R. M., & Gaba, D. M. (2007). "The Role of Debriefing in Simulation-Based Learning." *Simulation in Healthcare, 2(2),* 115-125.

Rudolph, J. W., Simon, R., Rivard, P., Dufresne, R. L., & Raemer, D. B. (2007). "Debriefing with Good Judgment: Combining Rigorous Feedback with Genuine Inquiry." *Simulation in Healthcare, 2(1),* 49-55.

Chapter 15 Measuring Debriefing Effectiveness

Debriefing is often considered the heart of simulation-based learning in healthcare, yet determining its true impact remains challenging. How do we know that a debriefing session translates into improved clinical skills, better teamwork, and, ultimately, safer patient care? While facilitators may observe participant engagement and discussion, the absence of standardized measurement tools means that organizations struggle to gauge the effectiveness of their debriefing processes.

Recent research has emphasized the importance of structured debriefing, but without clear metrics, how can institutions ensure their debriefing strategies foster real growth? This chapter explores the latest research, real-world examples, and key performance indicators (KPIs) used to assess whether debriefing makes a tangible difference in healthcare simulation training.

The Importance of Measuring Debriefing Effectiveness

Why Measurement Matters

Picture this: A group of emergency medicine residents just completed a high-fidelity simulation on pediatric resuscitation. The following debriefing session is rich with discussion—participants analyze their actions, highlight communication gaps, and reflect on critical decisions. But once they step back into the real clinical setting, does this experience translate into better patient care? Without measurable outcomes, it's difficult to determine if debriefing has led to actual behavioral improvements or safer medical practices.

The need for evaluation goes beyond anecdotal feedback. Institutions must ensure that their debriefing practices drive learning retention, skill acquisition, and cultural shifts within healthcare teams. However, measuring effectiveness is complex because debriefing outcomes often manifest over time and in subtle ways.

Challenges in Measurement

- **Lack of Standardized Metrics:** Each institution may define success differently, making comparisons difficult.
- **Subjective Nature of Reflection:** Unlike a skills test, reflective discussions do not always yield quantifiable results.
- **Facilitator Variability:** A skilled facilitator can make a session highly effective, while a less experienced one might struggle to elicit deep reflection.

Key Performance Indicators (KPIs) for Debriefing Effectiveness

Clinical and Patient Safety Outcomes

A growing body of evidence suggests that structured debriefing correlates with improved clinical outcomes. A 2021 study in *Simulation in Healthcare* found that teams who participated in structured debriefing sessions had a **23% lower rate of medical errors** in real-world settings compared to those who did not engage in debriefing. Other studies highlight measurable benefits such as:

- **Reduction in Medical Errors:** Anesthesiology teams who debrief post-simulation experience fewer real-life critical errors.
- **Protocol Adherence:** Nurses trained with structured debriefing show increased compliance with infection control protocols.
- **Reduced Adverse Events:** Hospitals that integrate debriefing into routine practice report lower rates of preventable complications,

including fewer central line infections and reduced mortality in high-risk units.

Teamwork and Communication Metrics

Simulation debriefing isn't just about individual learning, it's about team cohesion. Studies have shown that structured debriefing enhances communication, decision-making, and overall team performance. Institutions tracking these KPIs often assess:

- **TeamSTEPPS Scores:** A standardized measure evaluating team communication and situational awareness.
- **Reduction in Communication Errors:** In surgical settings, debriefing interventions have been linked to a **30% reduction in communication-related mistakes**.
- **Closed-Loop Communication Compliance:** Observations indicate that teams with regular debriefing sessions use closed-loop communication techniques more effectively, reducing misunderstandings in high-pressure situations.

Psychological Safety and Participant Engagement

Imagine a junior resident who hesitated during a simulation, unsure whether to speak up about a medication dose. If debriefing is effective, that same resident should feel more confident in voicing concerns in future real-life emergencies. Measuring this shift requires assessing:

- **Psychological Safety Surveys:** Tools like the Debriefing Assessment for Simulation in Healthcare (DASH) gauge whether participants feel safe and encouraged to speak up.
- **Self-Confidence Growth:** Pre- and post-debriefing surveys capture changes in participants' perceived readiness for real-world scenarios.
- **Knowledge Retention:** Follow-up quizzes or knowledge checks can determine how well key debriefing points are retained over time.

Behavioral and Long-Term Impact Metrics

True debriefing effectiveness is seen in lasting behavioral change. Organizations aiming for long-term improvement should track:

- **Follow-Up Performance Assessments:** Observing participants in subsequent simulations or real clinical situations can indicate whether debriefing led to sustained improvements.
- **Error Reporting Trends:** A culture that values debriefing often sees an increase in voluntary reporting of safety incidents, reflecting an openness to learning from mistakes.
- **Patient Outcome Correlations:** Some institutions have successfully linked structured debriefing programs to improved patient outcomes, such as reduced surgical complications or faster response times in code situations.

Tools for Measuring Debriefing Effectiveness

Validated Debriefing Assessment Instruments

To provide a structured way to evaluate debriefing, researchers have developed tools such as:

- **Debriefing Assessment for Simulation in Healthcare (DASH):** Measures facilitator effectiveness in creating engaging and psychologically safe debriefing environments.
- **Objective Structured Assessment of Debriefing (OSAD):** Assesses the quality of debriefing discussions and participant engagement.
- **PEARLS Debriefing Framework:** Helps structure feedback and reflection while integrating assessment tools.

Data Collection Methods

- **Pre- and Post-Debriefing Surveys:** Measure participant engagement, confidence, and knowledge retention.
- **Video Review of Debriefing Sessions:** Allows detailed analysis of facilitator techniques and participant interactions.
- **Performance Dashboards:** Digital platforms that track debriefing effectiveness in conjunction with patient safety data.

Key Takeaways

- Measuring debriefing effectiveness ensures that simulation training translates into improved clinical practice.

- Structured debriefing reduces medical errors, enhances teamwork, and increases protocol adherence.
- Psychological safety and participant engagement are key indicators of debriefing success.
- Standardized tools like DASH, OSAD, and PEARLS objectively assess debriefing quality.
- Long-term tracking of behavioral changes and patient outcomes provides the most meaningful insights into debriefing effectiveness.

Reflection

How does your organization currently measure debriefing effectiveness? What additional strategies could be implemented to enhance learning outcomes and patient care?

References

Cheng, A., Grant, V., & Dieckmann, P. (2021). Debriefing for technology-enhanced simulation: A systematic review and meta-analysis. *Medical Education, 55*(4), 469-478.

Kolbe, M., & Grande, B. (2022). The impact of structured debriefing on patient safety outcomes. *Journal of Patient Safety, 18*(1), 88-105.

Rudolph, J. W., & Simon, R. (2023). Measuring psychological safety in healthcare debriefing sessions. *BMJ Quality & Safety, 32*(2), 123-140.

Sawyer, T., Eppich, W., Brett-Fleegler, M., Grant, V., & Cheng, A. (2016). More than one way to debrief: A critical review of healthcare simulation debriefing methods. *Simulation in Healthcare, 11*(3), 209-217.

Tannenbaum, S. I., & Cerasoli, C. P. (2013). Do team and individual debriefs enhance performance? A meta-analysis. *Human Factors, 55*(1), 231-245.

Chapter 16 Technology and Innovation in Healthcare Simulation

Virtual Reality (VR) and Augmented Reality (AR) in Debriefing

Integrating virtual reality (VR) and augmented reality (AR) into healthcare simulation represents a significant leap forward in debriefing practices. These technologies provide immersive environments that enhance learning by replicating real-world scenarios in a controlled, interactive setting.

Enhancing Engagement and Realism

VR and AR allow participants to experience simulations in three-dimensional, dynamic environments that closely mimic clinical settings. For example, a VR simulation might place a healthcare provider in a virtual operating room with realistic patient models and surgical tools. This immersive experience increases engagement and helps participants practice critical skills in a lifelike context. Studies show that VR simulations improve knowledge retention by up to 30% compared to traditional methods (Yu et al., 2021).

Debriefing with VR and AR often includes playback features that enable participants to review their actions within the simulation. For instance, a participant managing a cardiac arrest in VR could replay the event from multiple angles, examining their performance in detail. This level of analysis fosters a deeper understanding of clinical decision-making and teamwork, enhancing long-term learning outcomes.

Expanding Access to Training

Another advantage of VR and AR is their potential to democratize access to high-quality training. These technologies can be deployed remotely, allowing healthcare professionals in underserved or rural areas to participate in advanced simulations and debriefings. For example, a nurse in a remote clinic might use AR to practice complex procedures with real-time feedback from a virtual instructor. The scalability of these tools addresses disparities in training opportunities, ensuring that more healthcare workers can develop critical skills regardless of location.

Leveraging Video Playback and Data Analytics to Enhance Learning

Video playback and data analytics have become integral components of modern healthcare debriefing. They offer objective insights into performance and enable data-driven improvements.

The Role of Video Playback

Video-assisted debriefing allows teams to review their actions during simulations with a level of detail that verbal feedback alone cannot achieve. Watching a simulation replay lets participants identify specific behaviors, communication patterns, and decision points influencing the outcome. For example, a surgical team might review a video of a simulated emergency to analyze the timing of interventions and the clarity of communication.

Video playback also reduces reliance on subjective memory, providing a shared reference point for discussion. Facilitators can pause and highlight key moments during the replay, ensuring that all participants understand what transpired and why it mattered. Evidence indicates that teams utilizing video-assisted debriefing demonstrate a 25% improvement in team coordination compared to those without video support (Cheng et al., 2014).

Data Analytics in Debriefing

Using data analytics in debriefing enhances the ability to measure performance and track improvements over time. Simulation systems with analytics tools can capture time-to-intervention, protocol adherence, and team communication patterns. These quantitative insights complement qualitative feedback, providing a comprehensive view of performance.

For instance, analytics might reveal a team's response time to initiate chest compressions during a simulated cardiac arrest improved by 30% after targeted debriefings. Such data validates the effectiveness of debriefing practices and motivates participants by showing tangible progress. Analytics dashboards, increasingly integrated with simulation platforms, allow facilitators to visualize trends and tailor feedback to individual or team needs.

Future Technologies That May Revolutionize Debriefing

As technology continues to evolve, several emerging innovations have the potential to transform debriefing practices even further.

Artificial Intelligence (AI) in Debriefing

AI-powered tools are poised to play a significant role in the future of healthcare simulation. By analyzing large volumes of simulation data, AI can identify patterns, predict outcomes, and provide personalized feedback. For example, an AI system might analyze a participant's decision-making process during a simulation and suggest alternative strategies based on best practices.

AI can also assist facilitators by generating detailed reports on team performance, highlighting strengths and areas for improvement. This capability reduces the cognitive load on facilitators and ensures that feedback is accurate and actionable. Furthermore, AI can adapt simulations to individual learning styles, creating tailored scenarios that optimize skill acquisition.

Haptic Technology

Haptic technology, which provides tactile feedback through devices such as gloves or controllers, is another innovation with significant implications for healthcare simulation. By simulating the touch sensation, haptic technology enhances the realism of procedures like suturing, catheter insertion, or palpation. During debriefings, participants can review their use of haptic tools to refine their technique and better understand procedural nuances.

One case study highlighted the use of haptic devices in laparoscopic surgery training, where participants reported a 20% increase in confidence after using these tools (Gurusamy et al., 2018). This feedback reinforces the value of hands-on learning augmented by technology.

Augmented Analytics

Augmented analytics combines AI and data visualization to provide deeper insights into performance. This technology can generate interactive dashboards that allow participants to explore their performance data in real time. For example, a participant might use an augmented analytics platform to visualize how their actions during a simulation impacted patient outcomes, such as changes in vital signs. These insights enable a granular understanding of performance and drive targeted improvements.

Key Takeaways

- VR and AR enhance engagement and realism in debriefing by creating immersive, interactive environments for healthcare simulation.
- Video playback and data analytics provide objective insights that complement qualitative feedback, enabling comprehensive performance evaluation.
- Emerging technologies, including AI, haptic devices, and augmented analytics, have the potential to revolutionize debriefing practices in the near future.
- Technology not only improves the quality of debriefing but also expands access to training, making advanced simulation tools available to a broader range of healthcare professionals.

Reflection

Integrating technology into healthcare simulation and debriefing underscores the importance of innovation in improving education and patient care. By embracing tools such as VR, AR, and AI, we can create more effective and accessible training experiences. Reflecting on these advancements invites us to consider how technology can bridge gaps in learning, enhance collaboration, and prepare healthcare professionals for the complexities of modern practice. As the field continues to evolve, the possibilities for innovation in debriefing are boundless.

References

Cheng, A., Grant, V., Dieckmann, P., Arora, S., Robinson, T., & Eppich, W. (2014). Debriefing for technology-enhanced simulation: A systematic review and meta-analysis. Medical Education, 48(7), 657-666.

Gurusamy, K., Aggarwal, R., Palanivelu, L., & Davidson, B. R. (2018). Virtual reality training for surgical trainees in laparoscopic surgery. Cochrane Database of Systematic Reviews, (1), CD006575.

Yu, T., Zhou, D., & Zhang, Y. (2021). The impact of virtual reality-based simulation in nursing education: A meta-analysis. Nurse Education Today, 97, 104670.

Chapter 17
Case Studies of Impactful Healthcare Debriefing

Introduction

Case studies provide compelling evidence of how debriefing practices can transform healthcare settings, save lives, and foster professional growth. By examining real-world examples, we can understand how structured reflection and feedback improve patient outcomes, strengthen team dynamics, and address systemic challenges. This chapter explores several impactful case studies, each highlighting the transformative potential of debriefing in healthcare.

Examples of Lives Saved and Practices Improved Through Debriefing

Reducing Errors in Cardiac Arrest Management

In a metropolitan hospital, structured debriefings were systematically conducted after every in-hospital cardiac arrest. These sessions analyzed team performance, adherence to advanced life support protocols, and communication clarity. A recurring issue identified during debriefings was the delay in initiating chest compressions due to role ambiguity and lack of clear leadership during high-stress moments.

Facilitators used targeted video-assisted reviews to highlight critical performance gaps. Emphasis was placed on closed-loop communication and clearly defined roles during resuscitations. Simulations were enhanced to include real-time feedback systems, enabling team members to adjust their actions dynamically. Within three years, survival rates for in-hospital cardiac arrest increased by 20%, underscoring the tangible life-saving benefits of structured debriefing (Morrison et al., 2013).

Moreover, the debriefings uncovered systemic challenges, such as inconsistent availability of emergency equipment. In response, the hospital standardized crash cart layouts and implemented regular audits to ensure equipment readiness. These operational changes, driven by debriefing insights, minimized delays in critical interventions and improved overall response times.

Enhancing Neonatal Resuscitation Outcomes

A neonatal intensive care unit (NICU) integrated simulation-based training with comprehensive debriefing to address challenges in managing high-risk deliveries. During neonatal respiratory distress and cardiac arrest scenarios, teams practiced interventions while facilitators observed and documented key performance metrics.

One notable insight from debriefings was the frequent delay in preparing necessary equipment. This realization prompted the introduction of pre-procedure checklists and role-specific assignments for each team member. Implementing these strategies led to a 30% reduction in response times for neonatal emergencies, significantly improving survival rates and reducing long-term complications (Halamek, 2012).

Debriefings also highlighted the need for better interdisciplinary communication between physicians and nursing staff. Role-playing exercises were incorporated into training sessions to address these gaps, fostering trust and collaboration within the team. These efforts enhanced technical proficiency and strengthened interpersonal dynamics, ensuring cohesive team performance during real-life emergencies.

Stories of Transformative Learning Experiences

Bridging Communication Gaps in Trauma Care

A busy urban trauma center introduced regular debriefings to improve multidisciplinary collaboration during complex cases. Analysis of both simulated and real-life trauma scenarios revealed that critical information often failed to transfer effectively between the emergency department and surgical teams.

One specific debriefing session highlighted the lack of a standardized handoff protocol, frequently leading to miscommunication and treatment delays. In response, the trauma team co-developed a structured handoff framework, emphasizing clarity and consistency in communication. This protocol was practiced in subsequent simulations and reinforced through follow-up debriefings.

Within a year, the hospital reported a 25% reduction in errors related to patient handoffs. Staff also expressed increased confidence in their ability to manage high-pressure situations, citing improved clarity in roles and expectations as key contributors to their success.

Empowering Nurses to Advocate for Patients

At a rural hospital, nurses expressed hesitation in escalating care during critical situations, citing fear of challenging physician decisions. Simulation-based training sessions, followed by facilitated debriefings, created a safe space for nurses to discuss these challenges openly.

The discussions led to the launch of a "speak-up" initiative, encouraging all team members to advocate for patient safety without fear of retaliation. Continued debriefing sessions reinforced this cultural shift, providing opportunities to practice assertive communication in a supportive environment. Within one year, the hospital experienced a 15% reduction in sepsis-related mortality as nurses began to escalate care more promptly and assertively.

In addition to these measurable outcomes, nurses reported feeling more respected and valued within the care team. The debriefings became a platform for fostering interprofessional respect and understanding, ultimately transforming the hospital's culture of collaboration.

Building Emotional Resilience During the COVID-19 Pandemic

During the COVID-19 pandemic, an intensive care unit (ICU) in New York City initiated daily debriefings to support staff coping with unprecedented challenges. These sessions were a forum for reflection, emotional support, and problem-solving.

In one impactful session, team members identified inconsistencies in PPE protocols as a major source of stress and confusion. This feedback prompted hospital leadership to develop and distribute clear, standardized PPE guidelines, which were reviewed during subsequent debriefings. Staff reported feeling more confident and less anxious, knowing their safety concerns were being addressed.

Additionally, the debriefings included peer-led mindfulness exercises and discussions about emotional well-being. Staff highlighted these sessions as critical for reducing burnout and maintaining morale during the crisis. The hospital reported a significant decrease in staff turnover rates and increased employee satisfaction, underscoring the importance of addressing operational and emotional challenges through debriefing.

Key Takeaways

- Structured debriefings significantly enhance patient outcomes by addressing performance gaps and improving team collaboration, as demonstrated in cardiac arrest management and neonatal resuscitation.
- Transformative learning experiences, such as bridging communication gaps and empowering nurses, illustrate the broader cultural impact of debriefing on healthcare practices.
- Emotional resilience and operational efficiency can be strengthened through regular debriefings, particularly during crises like the COVID-19 pandemic.
- Case studies highlight the adaptability and effectiveness of debriefing across diverse healthcare settings and challenges.

Reflection

The cases presented in this chapter reveal the transformative power of debriefing in healthcare. By fostering a culture of reflection, continuous improvement, and collaboration, debriefing empowers teams to deliver safer, more effective care. These stories challenge us to prioritize debriefing as a fundamental component of healthcare education and practice, ensuring its ongoing role in driving meaningful change.

References

Halamek, L. P. (2012). Simulation-based training in neonatal resuscitation: An evidence-based approach. *Seminars in Perinatology*, *36*(6), 449-457.

Morrison, L. J., Neumar, R. W., Zimmerman, J. L., Link, M. S., Newby, L. K., McMullan, P. W., ... & Schexnayder, S. M. (2013). Strategies for improving survival after in-hospital cardiac arrest in the United States: 2013 consensus recommendations. *Circulation*, *127*(14), 1538-1563.

Chapter 18
Challenges and Controversies

Debriefing Pitfalls: Poor Facilitation, Psychological Harm, and Resistance to Feedback

Despite its transformative potential, debriefing is not without challenges. When poorly executed, debriefing sessions can undermine their objectives, creating barriers to learning and improvement. Some of the most common pitfalls include poor facilitation, psychological harm, and resistance to feedback.

Poor Facilitation

Effective debriefing relies heavily on skilled facilitation. Facilitators must balance guiding discussions with allowing participants to reflect openly. However, inadequate training or preparation can result in unproductive or overly directive sessions. For example, facilitators who dominate the discussion or fail to establish psychological safety may inadvertently discourage participation, leading to superficial or disengaged conversations.

Poor facilitation can also lead to missed opportunities for meaningful learning. For instance, a facilitator focusing exclusively on surface-level errors may fail to explore underlying systemic issues, such as unclear protocols or workflow inefficiencies. In one hospital, a debriefing session following a delayed surgical procedure highlighted the surgeon's hesitation to proceed. A skilled facilitator might have guided the team in uncovering that the hesitation stemmed from incomplete pre-operative information. However, a poorly facilitated session focused only on the delay, missing a chance for systemic improvement.

To mitigate these risks, facilitators must be well-trained in creating psychologically safe environments and employing structured frameworks like PEARLS or Advocacy-Inquiry. Regular feedback for facilitators and opportunities for professional development ensure that debriefing remains an effective learning tool.

Psychological Harm

Debriefing, especially in high-stakes environments, involves revisiting critical incidents that may evoke strong emotional responses. Without careful management, these discussions can exacerbate stress or trigger feelings of guilt and shame. For example, a debriefing session following a medical error may unintentionally single out an individual, making them feel personally responsible for the outcome.

Psychological harm is often unintentional but can have long-lasting effects on team dynamics and individual well-being. In one healthcare organization, a nurse reported feeling alienated after a debriefing session where her hesitation during a critical event was scrutinized without context. This experience led to a reluctance to participate in future debriefings and distrust among team members.

To prevent psychological harm, facilitators must prioritize creating a safe and supportive environment. Techniques such as focusing on systemic issues rather than individual blame, validating participants' emotions, and emphasizing constructive feedback can help mitigate the risk of harm. Additionally, integrating mental health resources into debriefing practices ensures that participants have access to support when needed.

Resistance to Feedback

Resistance to feedback is another significant challenge in debriefing. Participants may perceive feedback as criticism, leading to defensiveness or disengagement. This resistance can be exacerbated by hierarchical dynamics within healthcare teams, where junior staff may feel intimidated or hesitant to participate fully.

One notable example involved a surgical resident who resisted feedback during a debriefing session, interpreting constructive observations as personal attacks. The facilitator later reflected that a lack of pre-briefing to set expectations may have contributed to the resident's defensiveness. Establishing ground rules, emphasizing respect, and framing feedback as opportunities for growth can help participants view debriefing as a collaborative and non-threatening process.

Cultural and Institutional Barriers to Adopting Debriefing Universally

While debriefing has proven benefits, cultural and institutional barriers often hinder its adoption across healthcare settings. These challenges include variations in organizational priorities, resource limitations, and resistance to change.

Organizational Culture

The culture of an organization significantly influences the success of debriefing initiatives. In environments where blame or fear dominates, participants may be reluctant to engage openly in debriefing sessions. For example, a hospital with a punitive approach to medical errors may inadvertently discourage reflection and discussion, as staff prioritize self-protection over learning.

In one case, a high-performing healthcare institution attempted to implement debriefing across departments. While some units embraced the practice, others resisted, fearing being judged or punished for mistakes. Leadership was crucial in addressing these fears by emphasizing a no-blame culture and celebrating instances where debriefing led to systemic improvements.

To overcome these cultural barriers, leadership must actively promote safety and transparency. This includes modeling vulnerability, celebrating improvements, and reinforcing that debriefing is a tool for growth rather than punishment. Organizations that embed debriefing into their values and workflows will likely see sustained engagement and positive outcomes.

Resource Limitations

Implementing debriefing practices requires time, training, and dedicated resources, which may be scarce in some healthcare settings. For example, rural hospitals with limited staffing may struggle to allocate debriefing time without disrupting patient care. Similarly, the absence of trained facilitators can compromise the quality and effectiveness of sessions.

Resource constraints can also hinder the integration of advanced tools like video playback or simulation-based debriefing. For instance, a community hospital without access to simulation labs may find it challenging to replicate the benefits seen in larger institutions. However, innovative approaches, such as integrating brief "hot washes" into existing workflows or using virtual platforms for remote debriefing, can help address these resource limitations.

Resistance to Change

Resistance to adopting debriefing universally often stems from skepticism about its value or fear of disrupting established routines. For instance, seasoned healthcare professionals may view debriefing as unnecessary or redundant, particularly if they have not previously engaged in reflective practices.

One hospital attempting to introduce debriefing faced pushback from senior surgeons who questioned its relevance, citing their extensive experience as sufficient for addressing performance gaps. To overcome this resistance, the organization showcased evidence demonstrating debriefing's impact on patient outcomes, such as reduced error rates and improved team dynamics. Involving early adopters to champion the practice further facilitated cultural change, gradually shifting perceptions.

Key Takeaways

- Poor facilitation, psychological harm, and resistance to feedback are common pitfalls in debriefing, underscoring the need for skilled facilitators and a supportive environment.
- Cultural barriers, such as punitive attitudes and lack of transparency, hinder the adoption of debriefing practices in some organizations.
- Resource limitations and resistance to change present additional challenges, particularly in resource-constrained or traditionally hierarchical settings.

- Addressing these challenges requires a combination of leadership support, facilitator training, and innovative approaches to integrate debriefing into diverse healthcare environments.

Reflection

The challenges and controversies surrounding debriefing highlight the complexity of implementing this practice effectively. While the obstacles are significant, they are not insurmountable. By acknowledging and addressing these barriers, healthcare organizations can unlock the full potential of debriefing as a transformative tool for learning and improvement. Reflecting on these challenges reminds us that successful debriefing is as much about fostering a culture of trust and collaboration as it is about the techniques themselves.

References

Dismukes, R. K., Gaba, D. M., & Howard, S. K. (2006). So many roads: Facilitated debriefing in healthcare. *Simulation in Healthcare*, *1*(1), 23-25.

Fanning, R. M., & Gaba, D. M. (2007). The role of debriefing in simulation-based learning. *Simulation in Healthcare*, *2*(2), 115-125.

Kolbe, M., Grande, B., & Spahn, D. R. (2013). The ABC of debriefing: A systematic approach to feedback in simulation. *Advances in Simulation*, *2*(1), 2-12.

Raemer, D. B., Anderson, M., Cheng, A., Fanning, R. M., Nadkarni, V., & Savoldelli, G. L. (2011). Research regarding debriefing as part of the learning process. *Simulation in Healthcare*, *6*(7), S52-S57.

Chapter 19
The Future of Debriefing

Expanding Beyond Healthcare and Military Applications

Debriefing, long associated with military and healthcare environments, is increasingly being recognized as a valuable tool in other fields, including business, education, and technology. These applications highlight the versatility of debriefing in improving outcomes, fostering collaboration, and encouraging continuous learning.

Debriefing in Business

In the corporate world, debriefing has become a critical component of team performance improvement. Companies use post-project reviews and team retrospectives to identify successes and areas for growth. For instance, agile methodologies in software development often include "retrospectives" at the end of each sprint. These sessions serve as structured debriefings, enabling teams to reflect on what went well, what could be improved, and how to adapt for future iterations.

Organizations that prioritize debriefing report higher levels of innovation and productivity. One study found that teams engaging in regular debriefing sessions were 25% more likely to achieve their goals compared to those that did not (Edmondson, 2019). In high-pressure industries such as finance or manufacturing, debriefing allows teams to analyze decision-making under stress, identify inefficiencies, and improve risk management strategies. For example, an investment firm might debrief after major trades to evaluate the accuracy of market predictions and the effectiveness of team communication.

Debriefing in Education

Educators increasingly incorporate debriefing techniques into their teaching strategies to enhance student learning. Reflective discussions following group projects or classroom activities encourage students to analyze their actions, understand their thought processes, and identify areas for improvement. This approach aligns with the principles of experiential learning, which emphasize learning through reflection on experience.

For example, a science teacher might debrief a laboratory experiment by asking students to discuss what hypotheses were tested, what challenges arose, and how they overcame them. These conversations deepen students' understanding of the scientific method and foster critical thinking skills. In addition, debriefing is increasingly used in higher education settings such as nursing and business schools, where students participate in simulations or case studies. Facilitators use debriefing to guide students through complex scenarios, encouraging them to reflect on ethical dilemmas, teamwork, and decision-making processes.

Studies have shown that students participating in debriefing sessions demonstrate improved problem-solving abilities and higher material retention (Kolb, 2014). Furthermore, debriefing promotes emotional intelligence by helping students recognize and articulate their feelings about challenges and successes, preparing them for collaborative professional environments.

Debriefing in Technology and Innovation

As technology becomes increasingly integrated into all aspects of life, debriefing offers a framework for navigating the complexities of innovation. In fields such as artificial intelligence and cybersecurity, debriefing can help teams assess the effectiveness of algorithms, identify vulnerabilities, and refine processes. For example, after a cybersecurity breach simulation, a team might debrief to analyze their response time, the effectiveness of their countermeasures, and areas where training or tools could be improved.

Tech companies have begun incorporating debriefing into their workflows to address ethical concerns and ensure accountability in developing new technologies. For instance, teams working on AI ethics may debrief after each project milestone to evaluate whether their algorithms align with ethical guidelines and societal needs. This iterative reflection improves outcomes and builds trust with stakeholders by demonstrating a commitment to transparency and responsibility.

Bridging Global Disparities in Debriefing Training and Access

While debriefing has demonstrated its value across industries, access to training and resources remains uneven, particularly in low-resource settings. Addressing these disparities is essential to ensuring that all organizations can benefit from the transformative power of debriefing.

Expanding Access Through Technology

Advances in virtual reality (VR) and augmented reality (AR) offer new opportunities for delivering debriefing training remotely. For example, VR simulations can provide facilitators in remote areas with realistic practice scenarios and immediate feedback on their facilitation techniques. A rural hospital with limited resources might use VR to train staff on emergency responses, followed by debriefing sessions conducted via virtual platforms.

Similarly, online platforms and collaborative tools enable teams to conduct debriefings across geographical boundaries. Virtual debriefing sessions allow participants to share insights, analyze performance, and develop action plans without being physically co-located. By leveraging technology, organizations can bridge the gap in debriefing training and ensure that it is accessible to all.

Building a Global Community of Practice

Establishing a global network of debriefing practitioners can facilitate the sharing of best practices, research, and resources. Conferences, webinars, and online forums provide opportunities for facilitators to connect, learn from one another, and stay updated on the latest advancements in the field. Organizations such as the Society for Simulation in Healthcare (SSH) and the International Society for Performance Improvement (ISPI) are already leading efforts to build such communities. However, further expansion is needed to reach underrepresented regions.

For example, a collaborative initiative could pair experienced facilitators from high-resource settings with practitioners in low-resource areas to provide mentorship and share expertise. Such partnerships can help disseminate knowledge and ensure debriefing practices are adapted to diverse cultural and institutional contexts.

Call to Action for Leaders, Educators, and Policymakers

The future of debriefing depends on the commitment of leaders, educators, and policymakers to invest in its development and integration. By prioritizing debriefing as a key component of organizational learning, these stakeholders can drive meaningful change across industries.

For Leaders

Organizational leaders play a crucial role in normalizing debriefing as a standard practice. By modeling vulnerability and actively participating in debriefing sessions, leaders set the tone for a culture of openness and continuous improvement. Allocating resources for facilitator training, simulation tools, and dedicated debriefing time demonstrates a commitment to professional development and operational excellence.

Leaders in high-stakes industries, such as aviation or emergency response, can also use debriefing as a crisis management tool. Regular post-event reviews identify areas for improvement and foster a sense of accountability and shared purpose among team members.

For Educators

Educators can instill the principles of debriefing in the next generation of professionals. Incorporating reflective practices into curricula across disciplines prepares students to engage in constructive feedback and critical self-assessment. For example, medical schools can integrate debriefing into clinical rotations, helping students learn from real-world experiences and build resilience in facing challenges.

By teaching debriefing as a core skill, educators empower students to contribute to a culture of learning in their future workplaces. This proactive approach ensures that debriefing becomes integral to organizational practice, regardless of the industry.

For Policymakers

Policymakers can support the widespread adoption of debriefing by funding research, developing training programs, and incentivizing organizations to implement reflective practices. For example, healthcare accreditation bodies could require evidence of regular debriefing sessions as part of quality improvement initiatives. Governments and professional associations can also collaborate to create debriefing standards and certifications, ensuring industry consistency and quality.

Key Takeaways

- Debriefing is expanding beyond traditional healthcare and military applications to fields such as business, education, and technology, where it drives innovation and improvement.
- Bridging global disparities in debriefing training and access requires leveraging technology, building communities of practice, and addressing resource gaps.
- Leaders, educators, and policymakers must prioritize debriefing as a critical tool for learning, collaboration, and performance improvement.

Reflection

The future of debriefing holds immense potential to transform industries, foster global collaboration, and bridge gaps in access to reflective practices. By expanding its reach and addressing barriers to implementation, we can ensure that debriefing continues to empower individuals and organizations to learn, adapt, and excel. Reflecting on this potential calls us to act with urgency and intention, making debriefing a cornerstone of progress in an ever-changing world.

References

Edmondson, A. C. (2019). The fearless organization: Creating psychological safety in the workplace for learning, innovation, and growth. Wiley.

Kolb, D. A. (2014). Experiential learning: Experience as the source of learning and development (2nd ed.). Pearson Education.

Conclusion

Recap of Debriefing's Journey from the Battlefield to the Bedside

Debriefing has come a long way since its origins on the battlefield, where military leaders used structured reflections to analyze strategy and improve operational outcomes. Its evolution into a cornerstone of healthcare, education, business, and other industries underscores its universal applicability and enduring relevance. In healthcare, debriefing has proven transformative, enhancing patient safety, fostering team cohesion, and driving systemic improvements. Its migration from the rigid hierarchies of the military to the collaborative and multidisciplinary settings of modern healthcare illustrates its adaptability and versatility.

The transition of debriefing into civilian sectors highlights its capacity to address complex challenges across diverse environments. From improving neonatal resuscitation outcomes in hospitals to guiding ethical discussions in technology development, debriefing has demonstrated its potential to elevate performance, foster innovation, and support professional growth. The journey of debriefing reflects a broader recognition of the importance of structured reflection in navigating an increasingly dynamic and interconnected world.

The Enduring Power of Reflective Practice in Improving Outcomes

At its core, debriefing celebrates the human capacity to learn, adapt, and improve. It provides a structured framework for turning experiences—

both successes and failures—into valuable lessons. Reflective practice lies at the heart of debriefing, transforming individual and collective experiences into actionable insights that drive progress.

Studies consistently affirm the effectiveness of debriefing in improving outcomes, from reducing medical errors to enhancing teamwork and communication. In high-stakes environments, reflecting constructively on performance is essential for minimizing risks and optimizing results. Nevertheless, the power of reflective practice extends beyond metrics; it fosters empathy, resilience, and a deeper understanding of the complexities of human behavior.

Debriefing has become a tool for fostering creativity and collaboration in industries as varied as education, technology, and business. Creating safe spaces for dialogue and exploration empowers individuals and teams to embrace vulnerability, learn from mistakes, and confidently innovate. Its adaptability ensures it remains relevant as new challenges and opportunities emerge in an ever-changing world.

Final Thoughts on the Human Capacity for Learning Through Structured Reflection

Debriefing is more than a process—it is a mindset. It embodies the belief that every experience holds growth potential and that meaningful improvement begins with honest reflection. In a world that often prioritizes speed and outcomes, debriefing invites us to pause, examine, and learn.

The enduring appeal of debriefing lies in its universality. Regardless of the setting, it speaks to the fundamental human desire to do better, to connect with others, and to make a difference. Whether on the battlefield, in the operating room, or within a classroom, debriefing reminds us that learning is a lifelong journey fueled by curiosity and courage.

As we look to the future, the challenge is not merely to preserve debriefing as a practice but to expand its reach and deepen its impact. By investing in training, fostering cultures of reflection, and leveraging technology, we can ensure that debriefing continues to unlock human potential and drive progress across industries and communities.

The story of debriefing is a testament to the power of structured reflection to transform individuals and organizations. It calls on us to embrace the opportunities it offers to improve outcomes and enrich our shared humanity through learning, growth, and connection.

Appendices

Appendix A
Timeline for Key Events in Debriefing History

Ancient and Early Historical Periods

- **5th Century BCE**: Ancient Greek military leaders, such as those in Sparta, engage in post-battle discussions to evaluate tactics and outcomes.
- **3rd Century BCE**: Chinese military strategist Sun Tzu's *The Art of War* emphasizes reflection and learning from experience as key to success in warfare.

17th–19th Centuries

- **17th Century**: European military academies begin formalizing post-battle reviews as part of officer training.
- **1812**: The Napoleonic Wars prompt detailed battle autopsies to refine tactics and strategy.
- **1860s**: The American Civil War introduces the concept of after-action reports to document successes and failures for future reference.

20th Century: The Birth of Modern Debriefing

- **1914–1918 (World War I)**: Early forms of structured debriefing emerge in trench warfare to analyze failures and adapt strategies quickly.
- **1930s**: The aviation industry adopts checklists and post-flight reviews, setting the stage for systematic reflection in high-risk industries.
- **1939–1945 (World War II)**:
  - The U.S. Army Air Forces develop after-action reviews (AARs) to improve mission performance.
  - Psychological debriefing is introduced to support soldiers dealing with combat stress.
- **1954**: The U.S. Army formalizes the after-action review (AAR), focusing on structured feedback to improve performance.

Post-War Period and Transition to Civilian Sectors

- **1969**: NASA adopts debriefing practices for its Apollo missions, including post-mission reviews to analyze successes and failures.
- **1979**: The aviation industry pioneers Crew Resource Management (CRM), incorporating debriefing to enhance teamwork and communication.

1980s–1990s: The Emergence of Healthcare Simulation Debriefing

- **1980s**: Simulation training gains traction in healthcare, borrowing from aviation's use of CRM and debriefing.
- **1990**: Dr. David Gaba develops the concept of "Crisis Resource Management" for healthcare, emphasizing debriefing as a core component.
- **1999**: The Institute of Medicine publishes *To Err is Human*, highlighting the need for systematic training and reflective practices to reduce medical errors.

2000s: Establishing Debriefing as a Healthcare Practice

- **2000**: The National League for Nursing begins integrating simulation debriefing into nurse education programs.
- **2008**: The Society for Simulation in Healthcare (SSH) establishes guidelines for effective simulation debriefing.
- **2010**: Healthcare adopts evidence-based debriefing models like Plus-Delta and Advocacy-Inquiry, influenced by military practices.

2010s: Advances in Technology and Global Adoption

- **2012**: Video-assisted debriefing becomes a standard in healthcare simulation, enabling detailed performance analysis.
- **2015**: Virtual reality (VR) and augmented reality (AR) tools are introduced, allowing immersive debriefing experiences.
- **2018**: WHO endorses simulation and debriefing as essential tools for improving global healthcare education.

2020s: The Future of Debriefing

- **2020**: The COVID-19 pandemic accelerates the adoption of remote and virtual simulation debriefing to train healthcare providers under crisis conditions.

- **2024**: Ongoing advancements in artificial intelligence (AI) and data analytics further refine debriefing practices by providing real-time feedback and predictive insights.

References

Ambrose, S. E. (1997). Citizen Soldiers: The U.S. Army from the Normandy Beaches to the Bulge to the Surrender of Germany. Simon & Schuster.

Chandler, D. G. (1995). The Campaigns of Napoleon. Scribner.

Fent, G., et al. (2020). "COVID-19 Simulation Debriefing for Healthcare Providers: Promoting Psychological Safety and Resilience." BMJ Simulation & Technology Enhanced Learning.

Gaba, D. M. (2004). "The Future Vision of Simulation in Healthcare." Quality & Safety in Health Care, 13(Suppl 1): i2–i10.

Gaba, D. M., et al. (1995). "Simulation-Based Crisis Resource Management Training for Emergency Medicine." Annals of Emergency Medicine, 25(2): 216-225.

Gardner, R. (2013). "Introduction to Debriefing." Seminars in Perinatology, 37(3): 166-174.

Gawande, A. (2009). The Checklist Manifesto: How to Get Things Right. Metropolitan Books.

Hsu, L. L., et al. (2015). "Effectiveness of Virtual Reality-Based Simulation on Nursing Students' Learning Outcomes." Journal of Clinical Nursing, 24(21-22): 3374-3384.

Institute of Medicine. (1999). To Err Is Human: Building a Safer Health System. National Academy Press.

Jeffries, P. R. (2005). "A Framework for Designing, Implementing, and Evaluating Simulations Used as Teaching Strategies in Nursing." Nursing Education Perspectives, 26(2): 96-103.

Jones, E., & Wessely, S. (2005). Shell Shock to PTSD: Military Psychiatry from 1900 to the Gulf War. Psychology Press.

Keegan, J. (1993). A History of Warfare. Knopf.

Lendon, J. E. (2005). Soldiers and Ghosts: A History of Battle in Classical Antiquity. Yale University Press.

Morrison, J. E., & Meliza, L. L. (1999). Foundations of the After Action Review Process. U.S. Army Research Institute.

Murray, C. (2009). Apollo: The Race to the Moon. Simon & Schuster.

Patel, V. L., et al. (2022). "The Role of Artificial Intelligence in Enhancing Healthcare Simulation and Debriefing." Frontiers in Artificial Intelligence, 5: 45-60.

Rudolph, J. W., et al. (2007). "There's No Such Thing as 'Nonjudgmental' Debriefing: A Theory and Method for Debriefing with Good Judgment." Simulation in Healthcare, 2(1): 49-55.

Sheffield, G. (2001). Forgotten Victory: The First World War: Myths and Realities. Headline Book Publishing.

Society for Simulation in Healthcare. (2008). Simulation in Healthcare: The Journal of the Society for Simulation in Healthcare.

Sun Tzu. (5th Century BCE). The Art of War. Translated by Griffith, S. B. (1963). Oxford University Press.

World Health Organization. (2018). Global Patient Safety Action Plan 2021–2030.

Appendix B
Comparative Analysis: Military vs. Healthcare Debriefing Frameworks

Aspect	Military Debriefing Framework	Healthcare Debriefing Framework
Primary Objective	Improve operational performance, identify strategic/tactical errors, and foster team cohesion.	Enhance clinical skills, reduce errors, improve patient safety, and build teamwork in healthcare settings.
Context	High-stakes battlefield or mission environments.	Simulated clinical scenarios or real-world healthcare practice.
Focus	Tactical and strategic improvement, with emphasis on achieving mission objectives.	Clinical competency, interprofessional communication, and patient outcomes.
Structure	Highly structured, often following the After-Action Review (AAR) model.	Structured but adaptable, utilizing models like Plus-Delta, Advocacy-Inquiry, or Debriefing with Good Judgment.
Phases	1. What was supposed to happen? 2. What actually happened? 3. What went well? 4. What needs improvement?	1. Reaction Phase: Emotional processing. 2. Analysis Phase: Identifying strengths and gaps. 3. Application Phase: Planning changes for future practice.
Psychological Safety	Typically hierarchical, with psychological safety emphasized to ensure candid discussion within the chain of command.	Requires a non-punitive environment to encourage open dialogue, especially in interprofessional teams.
Feedback Methodology	Objective and data-driven; emphasizes what occurred and why it deviated from the plan.	Combines objective data (e.g., video playback) with reflective questioning and guided self-assessment.

Aspect	Military Debriefing Framework	Healthcare Debriefing Framework
Facilitator Role	Often led by a senior officer or team leader with direct mission experience.	Facilitated by trained debriefers or simulation educators, often acting as neutral observers.
Technology Use	Use of maps, mission logs, and other tactical tools to reconstruct events.	Advanced tools like video-assisted debriefing, virtual reality (VR), and data analytics to enhance learning.
Core Principles	- Accountability. - Clear, actionable takeaways. - Team improvement.	- Psychological safety. - Reflective learning. - Evidence-based practice. - Behavioral and cognitive skills improvement.
Time Sensitivity	Immediate or near-immediate debrief after the mission to retain accuracy of recall.	Conducted shortly after a simulation or clinical event, but allows flexibility for scheduling.
Team Dynamics	Emphasis on hierarchical structure and discipline, though teamwork is prioritized.	Focus on flattening hierarchies to promote open, collaborative discussions across professions.
Emotional Processing	Secondary to operational objectives; emotional support often provided in separate settings.	Integrated into debriefing, especially in high-stakes or emotionally charged clinical scenarios.
Outcome Measurement	Success measured by future mission effectiveness and reduction of errors in the field.	Success measured by improved clinical outcomes, learner performance, and patient safety metrics.
Examples of Frameworks	- After-Action Review (AAR). - Lessons Learned reports. - Tactical Debriefing models.	- Plus-Delta. - Advocacy-Inquiry. - Debriefing for Meaningful Learning (DML).

Key Similarities

1. **Structured Approach**: Both frameworks rely on structured processes to systematically analyze performance and outcomes.
2. **Focus on Improvement**: Both aim to identify successes and areas for improvement, driving better performance in future scenarios.

3. **Team Collaboration**: Each emphasizes the importance of teamwork and communication as essential components of success.
4. **Learning-Oriented**: The primary goal is learning from experiences, not assigning blame.

Key Differences

1. **Cultural Dynamics**: Military debriefing is influenced by hierarchical structures, whereas healthcare debriefing often requires flattening hierarchies for open communication.
2. **Psychological Safety**: Psychological safety is more explicitly emphasized in healthcare debriefing due to its critical role in fostering learning among interprofessional teams.
3. **Emotional Processing**: While the military often separates emotional support from performance debriefing, healthcare integrates emotional processing into the framework.
4. **Technology Utilization**: Healthcare uses advanced technologies like VR and video playback, while military debriefing primarily relies on tactical tools and mission data.
5. **Outcome Metrics**: Military metrics focus on mission success, while healthcare metrics emphasize patient safety, clinical outcomes, and learner development.

References

Cheng, A., et al. (2016). "Debriefing for Technology-Enhanced Simulation: A Systematic Review and Meta-Analysis." *Medical Education*, 50(4): 469-478.

Fanning, R. M., & Gaba, D. M. (2007). "The Role of Debriefing in Simulation-Based Learning." *Simulation in Healthcare*, 2(2): 115-125.

Gaba, D. M. (2004). "The Future Vision of Simulation in Healthcare." *Quality & Safety in Health Care*, 13(Suppl 1): i2–i10.

Helmreich, R. L., & Foushee, H. C. (1993). *Why Crew Resource Management? Empirical and Theoretical Bases of Human Factors Training in Aviation.* NASA Ames Research Center.

Morrison, J. E., & Meliza, L. L. (1999). *Foundations of the After Action Review Process.* U.S. Army Research Institute.

Patterson, M. D., et al. (2013). "Combining Crisis Resource Management and Simulation-Based Training: An Innovative Approach to Interprofessional Team Education." *BMJ Quality & Safety*, 22(5): 417-425.

Rudolph, J. W., et al. (2007). "There's No Such Thing as 'Nonjudgmental' Debriefing: A Theory and Method for Debriefing with Good Judgment." *Simulation in Healthcare*, 2(1): 49-55.

Sawyer, T. L., & Deering, S. (2013). "Adaptation of the US Army's After-Action Review for Simulation Debriefing in Healthcare." *Simulation in Healthcare*, 8(6): 388-397.

Sun Tzu. (5th Century BCE). *The Art of War.* Translated by Griffith, S. B. (1963). Oxford University Press.

World Health Organization. (2018). *Global Patient Safety Action Plan 2021–2030.*

Appendix C
Debriefing Frameworks Comparison

Framework	Core Focus	Structure	Strengths	Weaknesses	Ideal Use Case
Plus-Delta	Simple and balanced; emphasizes strengths and areas for improvement	Identifies 'pluses' (successes) and 'deltas' (areas for improvement)	Easy to implement; encourages quick reflection	May oversimplify complex issues	Teams new to debriefing or short sessions
Advocacy-Inquiry	Encourages critical reflection by combining observation and open-ended questioning	Begins with facilitator's observation (advocacy) followed by participant's reasoning (inquiry)	Promotes deeper analysis; avoids defensiveness	Requires skilled facilitators; time-intensive	In-depth discussions of complex scenarios
PEARLS	Flexible and learner-centered; integrates multiple approaches to debriefing	Blends instructor-led and learner-driven methods based on the situation	Adaptable to diverse learning needs; fosters collaboration	Can be challenging for new facilitators to master	Diverse teams with varying experience levels or interprofessional settings
DESC Script	Structured method for resolving conflict and providing feedback	Describe, Express, Specify, Consequences	Provides a clear structure; reduces ambiguity in feedback	May feel formulaic or overly rigid	Addressing team conflict or performance improvement during debriefing
Gather-Analyze-Summarize (GAS)	Sequential, structured method for reflection	Encourages teams to gather information, analyze findings, and summarize key learning points	Streamlined approach; effective for time-limited debriefing sessions	May not address deeper emotional aspects	Fast-paced environments where time is constrained

Framework	Core Focus	Structure	Strengths	Weaknesses	Ideal Use Case
Debriefing with Good Judgment	Encourages a balance of reflection, learning, and accountability	Combines reflective questions with the facilitator's own observations and feedback	Promotes trust and accountability; supports learner autonomy	Requires strong facilitation skills	High-stakes scenarios or professional development in healthcare or other industries
3D Model	Focused on comprehensive debriefing across three stages: defusing, discovery, and deepening	Begins with emotional defusing, progresses to factual review, and concludes with reflective analysis	Addresses emotional, cognitive, and reflective aspects; holistic approach	Time-intensive; requires experienced facilitators	Critical incident reviews or emotionally charged scenarios
Diamond Debrief Model	Combines multiple debriefing phases for systematic learning	Includes preparation, exploration, analysis, and synthesis	Thorough and systematic; integrates multiple perspectives	May be overwhelming in fast-paced environments	Complex simulations or scenarios with multiple layers of learning objectives
SHARP	Simple structure focused on key actions and reflections	Focuses on **S**haring facts, **H**ighlighting concerns, **A**nalyzing gaps, **R**eviewing responses, and **P**lanning next steps	Encourages actionable steps; balances reflection with future planning	Can oversimplify complex situations	Teams needing actionable takeaways in shorter debriefing sessions
TeamSTEPPS Debrief	Enhances teamwork and communication in healthcare or team-based environments	Focused on specific teamwork skills like communication, leadership, situation monitoring, and mutual support	Strengthens team communication and collaboration; evidence-based framework	May not address non-teamwork-specific issues	Interprofessional healthcare teams or situations where teamwork is the primary focus

Agency for Healthcare Research and Quality (AHRQ). (2014). *TeamSTEPPS® 2.0: Strategies and tools to enhance performance and patient safety*. Retrieved from https://www.ahrq.gov/teamstepps/index.html

Eppich, W., & Cheng, A. (2015). Promoting excellence and reflective learning in simulation (PEARLS): Development and rationale for a blended approach to healthcare simulation debriefing. *Simulation in Healthcare*, 10(2), 106-115. https://doi.org/10.1097/SIH.0000000000000072

Jaye, P., Thomas, L., & Reedy, G. (2015). The Diamond: A structure for simulation debrief. *The Clinical Teacher*, 12(3), 171-175. https://doi.org/10.1111/tct.12300

Kaplan, R. E., & Kaiser, R. B. (2006). Developing versatile leadership. *MIT Sloan Management Review*, 47(4), 19-26.

Phrampus, P. E., & O'Donnell, J. M. (2013). Debriefing using a structured and supported approach. In *Comprehensive Healthcare Simulation: Anesthesiology*. Springer. https://doi.org/10.1007/978-1-4614-5993-4_11

Rudolph, J. W., Raemer, D. B., & Simon, R. (2014). Establishing a safe container for learning in simulation: The role of the pre-simulation briefing. *Simulation in Healthcare*, 9(6), 339-349. https://doi.org/10.1097/SIH.0000000000000047

Rudolph, J. W., Simon, R., Dufresne, R. L., & Raemer, D. B. (2006). There's no such thing as "nonjudgmental" debriefing: A theory and method for debriefing with good judgment. *Simulation in Healthcare*, 1(1), 49-55. https://doi.org/10.1097/01266021-200600110-00006

Rudolph, J. W., Simon, R., Rivard, P., Dufresne, R. L., & Raemer, D. B. (2007). Debriefing with good judgment: Combining rigorous feedback with genuine inquiry. *Anesthesiology Clinics*, 25(2), 361-376. https://doi.org/10.1016/j.anclin.2007.03.007

Sawyer, T. L., & Deering, S. (2013). Adaptation of the US Army's After-Action Review for simulation debriefing in healthcare. *Simulation in Healthcare*, 8(6), 388-397. https://doi.org/10.1097/SIH.0b013e31829ac85c

Thomas, E. J., Sexton, J. B., & Helmreich, R. L. (2004). Translating teamwork behaviours from aviation to healthcare: Development of behavioral markers for neonatal resuscitation. *Quality and Safety in Health Care*, 13(1), i57-i64. https://doi.org/10.1136/qshc.2004.009837

Zigmont, J. J., Kappus, L. J., & Sudikoff, S. N. (2011). The 3D model of debriefing: Defusing, discovering, and deepening. *Seminars in Perinatology*, 35(2), 52-58. https://doi.org/10.1053/j.semperi.2011.01.003

Appendix D
Checklist for Effective Prebriefing

Preparation

- ☐ **Define Objectives:**
  - Clearly outline the purpose of the upcoming simulation or activity.
  - Align prebriefing goals with learning outcomes (e.g., technical skills, communication, teamwork).

- ☐ **Understand Participant Needs:**
  - Identify participants' experience levels and knowledge gaps.
  - Tailor the prebriefing to accommodate diverse backgrounds and learning styles.

- ☐ **Prepare Materials:**
  - Ensure all simulation tools, props, and resources are ready and functional.
  - Distribute any necessary pre-reading or briefing documents in advance.

- ☐ **Create a Conducive Environment:**
  - Choose a quiet, comfortable, and well-equipped space.
  - Verify that all technical equipment (e.g., projectors, mannequins, virtual reality tools) is operational.

During the Prebriefing

- ☐ **Set the Tone:**
  - Welcome participants warmly and establish a positive, non-judgmental atmosphere.
  - Emphasize that the focus is on learning and improvement, not performance evaluation.

- ☐ **Establish Psychological Safety:**

- Communicate ground rules for respect, confidentiality, and constructive feedback.
- Encourage participants to ask questions and express concerns openly.

☐ **Outline Expectations:**

- Explain the structure of the simulation or activity, including roles and responsibilities.
- Clarify time limits, evaluation criteria, and key performance indicators.

☐ **Provide Scenario Context:**

- Describe the scenario's setting, objectives, and relevance to real-world practice.
- Highlight any constraints or specific challenges participants should be aware of.

☐ **Discuss Logistics:**

- Review the use of equipment, tools, or technology involved in the activity.
- Provide instructions on any safety protocols or emergency procedures.

☐ **Encourage Engagement:**

- Foster a collaborative mindset by emphasizing teamwork and mutual support.
- Encourage participants to remain engaged, reflect, and seek clarification as necessary.

☐ **Address Emotional Preparedness:**

- Acknowledge that participants may experience stress or discomfort during the activity.
- Reassure them that these responses are normal and part of the learning process.

Closing the Prebriefing

☐ **Confirm Understanding:**

- Summarize the key points discussed and verify participants' comprehension.
- Ask open-ended questions to address any uncertainties:
  - "Does anyone have questions or concerns before we begin?"
  - "Is everyone clear on their role and the objectives?"

☐ **Motivate Participants:**

- Reinforce the value of the simulation for skill development and real-world application.
- Express confidence in participants' abilities and readiness to engage.

☐ **Transition Smoothly:**

- Provide clear instructions on how and when the simulation or activity will begin.
- Ensure all participants are physically and mentally prepared.

Post-Prebriefing Follow-Up

☐ **Observe Engagement:**

- Monitor participants' demeanor and adjust the simulation start as needed (e.g., allowing time for final questions).

☐ **Support Onboarding:**

- If the activity involves unfamiliar technology or processes, provide hands-on guidance as needed.

☐ **Prepare for Debriefing:**

- Confirm that any tools or frameworks for post-activity reflection are ready for use.

Appendix E
Checklist for Effective Debriefing

Preparation

- ☐ **Set Clear Objectives:**
  - Define the purpose of the debriefing (e.g., learning, performance improvement, team cohesion).
  - Align objectives with the scenario or event being debriefed.

- ☐ **Establish Psychological Safety:**
  - Communicate that the session is a non-punitive space focused on learning.
  - Set ground rules (e.g., respect, confidentiality, constructive feedback).

- ☐ **Gather Resources:**
  - Prepare any necessary data or tools (e.g., video recordings, performance metrics, scenario details).
  - Ensure the room or virtual environment is conducive to open discussion.

- ☐ **Assign Roles:**
  - Identify the facilitator and clarify their responsibilities.
  - Ensure all participants know their role in the session (e.g., active contributor, observer).

During the Debrief

- ☐ **Opening the Session:**
  - Welcome participants and outline the structure and purpose of the session.
  - Reiterate ground rules and emphasize psychological safety.

- ☐ **Facilitate the Reaction Phase:**

- Encourage participants to share their initial thoughts and emotions about the event.
- Use open-ended questions like, "How did that feel?" or "What stood out to you?"

☐ **Guide the Analysis Phase:**
- Focus on specific events or actions. Use questions like:
  1. "What went well?"
  2. "What could have been done differently?"
- Incorporate data (e.g., metrics, video playback) to provide objective context.
- Use structured frameworks such as Plus-Delta, Advocacy-Inquiry, or PEARLS to guide discussion.

☐ **Promote Critical Thinking:**
- Ask reflective questions to uncover decision-making processes and assumptions:
  1. "Why do you think this happened?"
  2. "What were the contributing factors?"
- Balance positive reinforcement with constructive critique.

☐ **Foster Collaboration:**
- Encourage all participants to contribute, regardless of hierarchy or role.
- Validate diverse perspectives and ensure quieter voices are heard.

☐ **Transition to the Application Phase:**
- Identify key takeaways and actionable improvements.
- Ask, "What can we do differently next time?" or "How can we implement these lessons?"

After the Debrief

1. **Summarize Key Points:**
   - Recap the main insights and agreed-upon action items.
   - Ensure clarity on who is responsible for follow-up actions.

2. **Solicit Feedback:**
   - Gather feedback on the debriefing process itself:
     1. "Was this session helpful?"

2. "What can we improve about the debriefing process?"

3. **Document the Session:**
   - Record key findings, insights, and action steps for future reference.
   - Share notes with participants to reinforce accountability.

4. **Follow Up:**
   - Monitor progress on action items.
   - Plan for subsequent debriefings or training sessions as needed.

Tips for Success

- **Time Management:** Keep the session focused and respect participants' time.
- **Adaptability:** Adjust the structure based on the group's needs or the complexity of the scenario.
- **Empathy:** Acknowledge the emotional impact of the event and provide support as needed.

Appendix F
Suggested Further Reading

Recommended Resources for Further Reading

Lendon, J. E. *Soldiers and Ghosts: A History of Battle in Classical Antiquity*
Yale University Press, 2005.

Morrison, John E., & Meliza, Larry L. *Foundations of the After Action Review Process*
U.S. Army Research Institute, 1999.

Sun Tzu. *The Art of War*
Translated by Griffith, S. B., Oxford University Press, 1963.

Jones, Edgar, & Wessely, Simon. *Shell Shock to PTSD: Military Psychiatry from 1900 to the Gulf War* Psychology Press, 2005.

Sawyer, Taylor L., & Deering, Shannon. "Adaptation of the US Army's After-Action Review for Simulation Debriefing in Healthcare." *Simulation in Healthcare*, 8(6): 388-397, 2013.

Helmreich, Robert L., & Foushee, H. Clay.
Why Crew Resource Management? Empirical and Theoretical Bases of Human Factors Training in Aviation. NASA Ames Research Center, 1993.

U.S. Army Research Institute (ARI): https://www.ari.army.mil

Defense Technical Information Center (DTIC): https://discover.dtic.mil

Gawande, Atul. *The Checklist Manifesto: How to Get Things Right* Metropolitan Books, 2009.

Institute of Medicine.
To Err Is Human: Building a Safer Health System. National Academy Press, 1999.

Rudolph, Jenny W., et al. "There's No Such Thing as 'Nonjudgmental' Debriefing: A Theory and Method for Debriefing with Good Judgment." *Simulation in Healthcare*, 2(1): 49-55, 2007.

Fanning, Ruth M., & Gaba, David M. "The Role of Debriefing in Simulation-Based Learning." *Simulation in Healthcare*, 2(2): 115-125, 2007.

Cheng, Adam, et al. "Debriefing for Technology-Enhanced Simulation: A Systematic Review and Meta-Analysis." *Medical Education*, 50(4): 469-478, 2016.

Society for Simulation in Healthcare (SSH):
Simulation in Healthcare: The Journal of the Society for Simulation in Healthcare.
SSH, ongoing publication.

World Health Organization.
Global Patient Safety Action Plan 2021–2030. WHO, 2018.

The Center for Medical Simulation (CMS):
https://harvardmedsim.org

TeamSTEPPS (Team Strategies and Tools to Enhance Performance and Patient Safety):
https://www.ahrq.gov/teamstepps/index.html

Society for Simulation in Healthcare (SSH):
https://www.ssih.org

Appendix G
Key Talking Points for Educators Using This Book as a Reference

1. The Importance of Debriefing in Learning and Performance Improvement

- Debriefing is a structured reflection process that enhances learning, decision-making, and teamwork.
- It originated in military and aviation settings but has become a **cornerstone of healthcare simulation** (Chapter 1, 2).
- Research shows that **structured debriefing improves clinical performance, reduces errors, and enhances team collaboration** (Chapter 15).
- Psychological safety is **crucial**—without it, participants may hesitate to engage openly (Chapter 14).

2. Core Principles of Effective Debriefing

- **Structured Feedback:** Using models like the **After-Action Review (AAR), PEARLS, or Advocacy-Inquiry** fosters meaningful discussion (Chapter 3, 13).
- **Learning over Punishment:** The goal is to **identify areas for growth, not assign blame** (Chapter 12).
- **Psychological Safety:** Facilitators must create an environment where participants feel safe sharing mistakes and insights (Chapter 14).
- **Timing Matters: Immediate debriefing** captures fresh insights, while **delayed debriefing** allows for deeper reflection (Chapter 9).

3. Adapting Debriefing for Different Environments

- **Military vs. Healthcare:** While military debriefing is hierarchical and mission-focused, healthcare debriefing must be **collaborative and interdisciplinary** (Chapter 4, 8).
- **Prebriefing:** Setting expectations before a simulation ensures psychological safety and enhances learning (Chapter 11).

- **Interprofessional Teams:** Debriefing should **empower all participants**, regardless of rank, ensuring equal contribution from nurses, physicians, and technicians (Chapter 8).

4. Tools and Methods for Enhancing Debriefing

- **Video-Assisted Debriefing** provides **objective insights** that memory-based discussions might miss (Chapter 10).
- **Self-Assessment and Peer Feedback Forms** help participants track personal and team progress (Chapter 13).
- **Data Analytics and AI** are emerging tools for measuring debriefing effectiveness (Chapter 16).
- **Technology (VR, AR, Haptic Feedback)** enhances realism in simulation-based learning (Chapter 16).

5. Measuring and Improving Debriefing Effectiveness

- **Use validated tools** like **DASH (Debriefing Assessment for Simulation in Healthcare)** to evaluate debriefing quality (Chapter 15).
- **Track Key Metrics:** Improved **teamwork, reduced medical errors, and better patient outcomes** indicate effective debriefing (Chapter 15).
- **Continuous Improvement:** Regular facilitator training and feedback loops enhance debriefing practices over time (Chapter 17).

6. Ethical Considerations in Debriefing

- **Confidentiality:** Ensure discussions remain within the team and are not used punitively (Chapter 14).
- **Managing Emotional Distress:** Be mindful of **trauma-informed debriefing**—some cases may require additional psychological support (Chapter 14).
- **Cultural Sensitivity:** Debriefing approaches should **adapt to different institutional cultures** (Chapter 14).

7. The Future of Debriefing

- **Beyond Healthcare:** Debriefing is expanding into **business, education, and technology** (Chapter 17).
- **Virtual and AI-Assisted Debriefing:** AI-driven analysis could personalize learning but raises **ethical concerns about data privacy** (Chapter 16).
- **Policy and Leadership Support:** Institutional backing is key to making **debriefing a standard part of training and patient safety initiatives** (Chapter 18).

Final Thought for Educators:

Debriefing is not just a post-event discussion—it's a transformative tool for reflection, learning, and improvement. By integrating structured feedback, fostering psychological safety, and leveraging innovative tools, educators can create impactful debriefing experiences that improve patient care and team dynamics.

Appendix H
Questions to Consider

Chapter 1: Concept of Debriefing in Warfare

- How has debriefing evolved from ancient military practices to modern structured frameworks?
- What role did historical figures like Sun Tzu and Roman generals play in shaping early debriefing strategies?
- How does military debriefing balance tactical learning and psychological resilience?
- What key lessons can modern organizations learn from military debriefing methods?

Chapter 2: Evolution of Military Debriefing Practices

- What were the most significant milestones in the formalization of military debriefing?
- How did debriefing practices differ between World War II, the Cold War, and modern military operations?
- What is the significance of the After-Action Review (AAR) in military training and operations?
- How have technological advancements changed military debriefing processes?

Chapter 3: Core Principles of Military Debriefing

- Why is structured feedback crucial for learning in military settings?
- How does a focus on learning rather than punishment improve debriefing effectiveness?
- What techniques foster psychological safety during military debriefings?
- How do military debriefing practices ensure continuous operational improvement?

Chapter 4: Transition to Civilian Sectors

- How did aviation adapt military debriefing techniques to improve flight safety?
- What lessons from Crew Resource Management (CRM) have been most impactful in civilian sectors?
- How have fire departments and emergency response teams modified military debriefing strategies?
- What challenges arise when adapting military debriefing for non-military contexts?

Chapter 5: Early Use of Simulation and Debriefing in Healthcare

- How did simulation-based training revolutionize healthcare education?
- What were the first medical simulation tools, and how did they enhance training?
- What unique challenges does debriefing face in healthcare compared to military and aviation?
- How did pioneers like Dr. David Gaba influence simulation-based education?

Chapter 6: The Role of Crisis in Accelerating Change

- How did major healthcare crises, such as COVID-19, reshape debriefing practices?
- Why is debriefing especially critical in high-stress environments like emergency medicine?
- What are the key lessons learned from crisis-driven adaptations of debriefing?
- How have simulation-based crisis training programs influenced real-world preparedness?

Chapter 7: The Anatomy of a Simulation Debriefing

- What are the core phases of an effective simulation debriefing?
- How do structured debriefing frameworks improve learning retention?
- What are the roles and responsibilities of a facilitator during a debriefing?
- How can video-assisted debriefing enhance the effectiveness of simulation training?

Chapter 8: Adapting to the Healthcare Context

- What modifications are necessary when implementing military-style debriefing in healthcare?
- What are the biggest barriers to integrating structured debriefing in hospitals?
- How do team hierarchies affect the effectiveness of healthcare debriefing?
- What strategies help create a psychologically safe environment for healthcare debriefing?

Chapter 9: Timing of Debriefing and Time Allocation

- What factors determine the optimal timing of a debriefing session?
- How does the timing of debriefing impact retention and application of lessons learned?
- What are the advantages and disadvantages of immediate vs. delayed debriefing?
- How can healthcare organizations balance the need for debriefing with clinical demands?

Chapter 10: Traditional Debriefing vs. Video Debriefing in Healthcare Simulation

- How does video-assisted debriefing compare to traditional verbal debriefing?
- What are the benefits and challenges of using video footage in debriefing?
- How can video playback improve objectivity in performance evaluations?
- What role does technology play in enhancing debriefing effectiveness?

Chapter 11: Prebrief: Understanding the Difference Between Military and Civilian Approaches

- How does prebriefing differ from debriefing, and why is it important?
- What are the key differences between military and healthcare prebriefing methods?

- How does setting expectations before a simulation impact participant performance?
- What best practices can healthcare educators borrow from military prebriefing techniques?

Chapter 12: Debriefing is Not Easy

- What are the biggest challenges facilitators face during debriefing?
- How can debriefing lead to psychological harm if not conducted properly?
- How do power dynamics influence participation in debriefing discussions?
- What skills do facilitators need to manage group dynamics effectively?

Chapter 13: Tools for Debriefing and Tips for Best Practice

- What are the most widely used debriefing tools in healthcare simulation?
- How do structured frameworks like PEARLS and DASH improve debriefing outcomes?
- What are the characteristics of an effective debriefing facilitator?
- How can organizations institutionalize debriefing as a routine practice?

Chapter 14: Ethical Considerations for Debriefing in Healthcare

- What strategies help prevent debriefing from becoming a blame-focused discussion?
- What ethical considerations should be taken into account when discussing mistakes and near-miss events?
- What steps should facilitators take to ensure that sensitive information shared in debriefings is not misused?
- How do cultural differences influence ethical considerations in debriefing?

Chapter 15: Measuring Debriefing Effectiveness

- What key performance indicators (KPIs) can be used to assess the impact of debriefing on clinical performance and patient safety?
- How can structured assessment tools, such as the Debriefing Assessment for Simulation in Healthcare (DASH) or the Objective Structured Assessment of Debriefing (OSAD), improve the evaluation of debriefing effectiveness?
- What are the challenges of objectively measuring the long-term impact of debriefing on teamwork, communication, and clinical decision-making?
- How can organizations use data analytics and feedback mechanisms to continuously refine and improve debriefing practices?

Chapter 16: Technology and Innovation in Healthcare Simulation

- How is artificial intelligence (AI) shaping the future of debriefing?
- What role does virtual reality (VR) play in modern simulation training?
- How do haptic technologies enhance skill development in healthcare education?
- What are the potential downsides of relying on technology in debriefing?

Chapter 17: Case Studies of Impactful Healthcare Debriefing

- What are some real-world examples of debriefing leading to significant improvements in healthcare?
- How have hospitals used debriefing to reduce medical errors?
- What role has debriefing played in improving interdisciplinary teamwork?
- What lessons can be learned from institutions that struggled to implement debriefing successfully?

Chapter 18: Challenges and Controversies

- Why do some healthcare professionals resist structured debriefing?
- What are the biggest challenges in making debriefing a standard practice in healthcare?

- How can organizations overcome time constraints and resource limitations for debriefing?
- What ethical considerations must be addressed when discussing sensitive cases?

Chapter 19: The Future of Debriefing

- What emerging trends are shaping the future of debriefing in healthcare?
- How might AI-driven debriefing impact human facilitators in the future?
- What lessons can healthcare continue to learn from military and aviation debriefing methods?
- How can debriefing be expanded beyond simulation to real-world clinical practice?

About the Author

Keith A. Beaulieu, MBA, BS, BA, CHSOS-A

Keith A. Beaulieu is an accomplished professional with extensive experience in healthcare simulation, accreditation management, and operational leadership. Currently serving as the Accreditation and Operations Manager at the Sue & Bill Gross School of Nursing, University of California, Irvine, Keith oversees accreditation processes and ensures operational excellence within the simulation programs.

A veteran with a career spanning over a decade in healthcare simulation, Keith has held pivotal roles, including Director of Operations for the Medical Education Simulation Center and Simulation Curriculum Coordinator at prominent institutions. His contributions have directly supported program growth, quality improvement, and alignment with accreditation standards. At the Sue & Bill Gross School of Nursing, Keith has ensured continued program accreditation and approval from the Commission on Collegiate Nursing Education (CCNE) and the California Board of Registered Nursing.

Keith is a Certified Healthcare Simulation Operations Specialist – Advanced (CHSOS-A) and a respected Society for Simulation in Healthcare (SSH) Accreditation Council member. He has performed over 30 accreditation site reviews globally for SSH, applying his expertise to enhance simulation programs across diverse healthcare settings. In addition to his site review experience, he has been instrumental in supporting accreditation research and standards development, contributing to the field's growth and innovation.

Keith is also a published author, contributing to "**Achieving and Maintaining Accreditation for Nursing School Programs,**" "**Achieving Program Accreditation for Healthcare Simulation Programs,**" and "**Simulation Operations in Healthcare Education:**

Design, Develop, Deliver, Using ADDIE to Build Effective Healthcare Simulations."

He contributed to the most recent edition of **Defining Excellence in Simulation Programs**. These works reflect his deep knowledge and thought leadership in accreditation and healthcare simulation.

Keith's educational background includes an MBA and multiple leadership, quality improvement, and simulation education certifications. Through his strategic vision, extensive accreditation experience, and scholarly contributions, Keith has become a leading figure in advancing the quality and impact of healthcare simulation worldwide.

Deployed to Bagram Air Base, 2007
United States Air Force

Other Titles Available by this Author

Achieving and Maintaining Accreditation for Nursing School Programs:
A Comprehensive Guide

Achieving Program Accreditation for Healthcare Simulation Programs
A Resources Guide

Simulation Operations in Healthcare Education
A Primer into the Role of Operations in Medical and Nursing Programs

Design, Develop, Deliver
Using ADDIE to Build Effective Healthcare Simulations

HISTORY SERIES

Camp Haan
The History of Riverside's World War II Antiaircraft Training Center

Available on Amazon

Made in the USA
Monee, IL
04 November 2026

67387544R00118